Perfumes & Perfumery
An Overview

Raymond A. Young

Copyright 2021
Self-Published

Cover Artwork by
Alvina Fredrickson

"Pleasure is the Flower that passes;
Remembrance, the lasting Perfume"
Jean De Boufflers

"Perfumery is the best imitation of the
Vibrancy and Subtlety of Nature's Evolution"
Marian Bendeth
Sixth Sense

To Kathryn
For her continued Love and Support
(even though she does not wear perfume)

Contents

Prologue	6
Chapter 1. Essential Oils and Odorants	7
Sources and Collection of Essential Oils	8
Essential Oils and Odorants from Plants	12
Odorants from Animal Sources	31
Synthetics	34
Chapter 2. Perfumes – History & Modern Perfumery	39
Brief History of Perfumes	39
Emergence of Modern Perfumery	41
Perfume Structure & Composition	42
Perfume Accords	45
Fragrance Perception	53
Timeline of Modern Perfumery	57
Celebrity Perfumes	73
Global Scale of Perfumery	75
Chapter 3. Creation of a Perfume	77
Olfactory Training Methods	77
Composing Perfumes	88
Computer Technology & Artificial Intelligence (AI) in Perfumery	98
Health and Safety Considerations	102
Chapter 4. Chemistry Odorants	106
Basic Chemistry of Odorants	106
Perfume Reactions and Stability	118
Gas Chromatography	119
Chemical Composition of Odorants	122
Chapter 5. Physiology and Theories of Smell	154
Physiology of Smell	154
Search for New Odorants	156
Theories of Smell	157
References and Bibliography	172
Appendix I. Odors of Some Perfumery Compounds	179
Appendix II. YouTube Lectures on Perfumes and Theories of Smell	185
About the Author	186
Subject Index	187

Prologue

I taught a number of different courses at the University of Wisconsin in Madison, from Chemistry of Natural Products to Ethnobotany. Some of the greatest student responses came when I passed around smell strips to the students while I discussed the properties of essential oils and the composition of perfumes. It was a superb way to garner their attention and I frequently noticed that the classroom soon took on a din of euphoria. The students were eager to gain further knowledge of fragrances and olfaction, so I further evaluated the substantial literature and books on perfumes and perfumery available on the market.

There are many excellent research papers and books available on essential oils, perfumes and perfumery. Many have a specific emphasis from essential oils and olfactive descriptions of perfumes to detailed chemistry of odorants. I also found that some of the best texts on the subject matter were very expensive, beyond the budget of many students. The intention of this treatise is therefore to provide a basic overview of the many important aspects related to the development of perfumes and to provide a fundamental understanding of the chemical basis of perfumery, all at reasonable price. The emphasis is on the development of fine fragrances.

This book is divided into five chapters: Essential Oils and Odorants, Perfumes - History and Modern Perfumery, Creation of a Perfume, Chemistry of Odorants and the Physiology and Theories of Smell. The first chapter covers the characteristics and isolation of some important essential oils from plant blossoms, leaves, roots & rhizomes, fruits, seeds, wood & bark, and plant exudates (resins). Also included in this chapter is a description of the characteristics of odorants from animal sources. Additional information about the chemical composition of many essential oils is covered in a later chapter on Chemistry of Odorants.

The basic structure and composition of perfumes is given in a separate chapter and more detailed descriptions are given for some of the historically important perfumes.

Methods of training for perfumery are reviewed and several approaches for composing new fragrances are described in the chapter on Creation of Perfumes. The increasing use of computer technology and artificial intelligence in perfumery is also described.

The chapter on the Chemistry of Odorants clarifies the structural character of the many aroma components utilized in perfumes. It is important to note at the onset that many of the chemicals derived from plants can be produced synthetically in the laboratory and many of the ingredients in perfumes on the market today are synthetic rather than plant derived.

The final chapter covers the Physiology of Smell and the complicated Theory of Smell. The molecular structure and vibrational theories of smell are described and conclusions are reached about the most probable mechanism of smell.

Chapter 1. Essential Oils and Odorants

Essential oils are pervasive in plant materials and provide the basic "essence" or fragrance of the plant component. These oils evaporate easily due to their high volatility or low boiling point providing the characteristic aromas (1-5).

Essential oils serve important functions for the plants. They attract insects and animals for both pollination and dispersal of seeds due to the fragrance of the flowers. Pollination is critical to agricultural production and it is estimated that bees pollinate 1/3 of all agricultural crops worth $17 billion annually. To the bee a flower is the fountain of life and to the flower a bee is a messenger of love!

In contrast, harmful bacteria and insect and animal depredation are repelled by essential oils. Resins produced by some trees deter wood boring beetles and the oils protect the plants against fungus and other microbes. Thus, plants use essential oils for both reproduction and protection.

The oils are composed of a wide range of chemicals such as terpenes, volatile phenols (such vanilla & clove scents) and aliphatic compounds. They are produced by many, but not all plants. The chemistry of essential oils is covered in detail in Chapter 4, The Chemistry of Odorants.

There are a wide variety of uses for essential oils. One of the most valuable economic uses is in the plethora of perfumes and cosmetics available on the market. Essential oils are also critical to the flavors of food and it has been estimated that 80% of taste is related to smell. Essential oils also have both antiseptic and bactericidal properties very valuable for therapeutic uses. Many native cultures use plant components for treatment of a wide range of aliments and this approach indirectly involves the use of essential oils for medicinal applications (1-5).

About two hundred plants are raised commercially for production of essential oils. As you will see in the following section, the processes for collection of essential oils from plants can be very time-consuming and expensive. This is exasperated by the very low yields of oil, often in the range of 0.1% to 1%, from the raw plant material. Thus, large amounts of the plant components must be collected and processed. The high cost is one of the reasons for the increasing use of synthetic chemicals in perfume formulations. However most high-end perfumes still maintain some fraction of the natural essential oils, which can be difficult to duplicate. The high cost also puts a burden on essential oil use for aromatherapy (1-5).

Odorants for perfumes have also been historically obtained from animal sources which is further described in this section. The use of synthetics in fragrances has increased dramatically in modern times and the use of these materials in perfume compositions is also reviewed in this section as well as in the chapter on Chemistry of Odorants.

Sources and Collection of Essential Oils

Essential oils can occur in all parts of plant materials including plant blossoms, leaves, woody material and plant exudates (resins). Table 1.1 summarizes the collection of natural essential oils and odorants and includes the part of the plant utilized for collection of the oil and the method of extraction, as further described below (1-16).

Methods of Collection of Essential Oils

There are a number of methods for collection of essential oils from plant materials and the choice depends on a number of variables, but mainly Cost and Quality of Product are the determining factors. The methods of collection are: Enfleurage, Solvent Extraction, Distillation, Expression and Tinctures & Resinoids. Please refer to the Table 1.1 which enumerates which extraction process and which plant component are utilized for collection of many natural fragrance materials (1-16). It should be noted that the strict definition of an "essential oil" is only for those oils obtained by distillation; however, the term is sometimes utilized for odorants obtained by the other methods.

Enfleurage

Enfleurage is a fascinating older method for collection of essential oils from flowers. It is hardly utilized today but an enlightening demonstration is given at the History of International Perfumes Museum in Grasse, France (museesdegrasse.com). It is a good choice for delicate flowers since the petals are hand placed on a plate of cold fat for several days which absorbs the perfume material to give a pomade. This waxy fat can also be heated for hot enfleurage which yields more essence. The pomade is then extracted with ethyl alcohol and the alcohol gently distilled away to give an enfleurage absolute. The method was traditionally utilized to obtain absolutes of rose, jasmine, violet and tuberose but is now rarely employed due to the high labor costs.

Distillation

Both hydro-distillation and steam distillation have been utilized for collection of essential oils, but steam distillation is the most commonly utilized method since yields can be significantly higher. Steam distillation is mainly used for blossoms and foliage but can be employed for any plant material. It is routinely utilized for extraction from, for example, lavender, orange blossoms, peppermint, patchouli, eucalyptus, etc. It usually involves a single process but the delicate ylang-ylang is purified by a fractional distillation process.

Table 1.1 Collection of Natural Fragrance Materials*

A-B
Ambrette (S 🝪), Angelica (R 🝪), Anise (S 🝪), Artemisia (AP 🝪), Balsam of Peru (E, boiling water), Basil (L 🝪), Bay (L 🝪), Bergamot (Fr ⚗), Benzoin (E 🝪), Birch tar (W 🝪), Black Currant buds (FL 🝪 🝪), Buchu leaf (L 🝪)

C
Cabrueva (W 🝪), Cade (W-Juniper 🝪), Cajeput (L,T 🝪), Calamus (R 🝪), Camphor (W 🝪), Caraway (S 🝪), Cardamom (S 🝪), Carnation^ (FL 🝪), Cassia (L 🝪), Cedarwood (W 🝪), Cedar Leaf (L 🝪), Celery (S 🝪), Chamomile (FL 🝪), Cinnamon bark (🝪), Cinnamon leaf (🝪), Citronella (L 🝪), Clary Sage (FL,L 🝪), Clove bud (🝪), Clove leaf (🝪), Copaiba balsam (E), Coriander (S 🝪), Cornmint (AP 🝪), Cumin (S 🝪 🝪)

D-F
Dill (AP 🝪), Elemi (E🝪 🝪), Eucalyptus (L 🝪), Fennel (S 🝪), Fir needle (🝪), Frankincense (Olibanum) (E, 🝪)

G-J
Galbanum (E 🝪🝪), Gardenia^ (FL 🝪), Geranium (L, stem 🝪 🝪), Ginger (R 🝪), Grapefruit (Fr ⚗), Guaiac wood (W 🝪), Ho (L,W 🝪), Hyacinth^ (FL 🝪), Jasmine (FL 🝪), Juniper Berry (Fr, 🝪 🝪)

L
Labdanum (Cistus) (E 🝪, AP 🝪), Lavender (AP 🝪), Lemon (Fr ⚗), Lemon Grass (L 🝪), Lilac^ (FL 🝪), Lime (Fr ⚗), Litsia cubeba (Fr ⚗),

M-O
Mandarin (Fr ⚗), Majoram (L,FL 🝪), Muget^ (FL 🝪), Myrrh (E 🝪), Neroli (FL 🝪), Nutmeg (Fr 🝪), Oakmoss (AP, 🝪), Opoponax (E 🝪), Orange (Fr, ⚗), Origanum (AP 🝪), Orris (Rhizome 🝪 🝪)

P
Patchouli (L 🝪), Penntroyal (AP, 🝪), Peppermint (AP 🝪), Petitgrain (L,T 🝪), Pimento (Fr, 🝪), Pine Oil (W, 🝪)

R-T
Rosemary (AP, 🝪), Rose (FL 🝪 🝪), Rosewood (🝪), Sage (AP 🝪), Sandalwood (W 🝪), Sassafras (R 🝪), Spearmint (AP 🝪), Star Anise (S 🝪), Storax (E, 🝪), Styrax (E 🝪), Tolu Oil (E, 🝪)

T-Y
Tangerine (Fr ⚗), Terragon (AP 🝪), Thyme (AP 🝪), Ti Tree (L 🝪), Tonka Bean (🝪), Tuberose (FL 🝪), Vanilla (Fr 🝪 🝪), Vetiver (R 🝪), Violet^ (FL 🝪), Ylang-Ylang (FL 🝪 🝪)

*Plant Part: AP=Aerial Parts, E=Exudate Resin, FL=Flower, Fr=Fruit, L=Leaves, R=Root, S=Seeds, T=Twigs, W=Wood; Method of Extraction: 🝪 = Distilled (Steam mostly or Dry), 🝪 = Solvent Extracted, ⚗ = Expression;
^Essential oils that are mostly synthetic recreations due to low yields and difficulty of extraction; References 1-14.

A diagram of a steam distilling apparatus is shown in Figure 1.1. The raw material is placed on a screen in a large vat and steam from heated water is passed through the plant material vaporizing the volatile compounds (with hydro-distillation the raw material is immersed in the water before heating). The vapors are passed through a condenser (cooled coil) which condenses the vapors back to a liquid which is collected in a receiving vessel. The water and oil are not miscible so the lower density oils separate out at the top of the vessel where they can be decanted.

Figure 1.1 Steam distillation apparatus

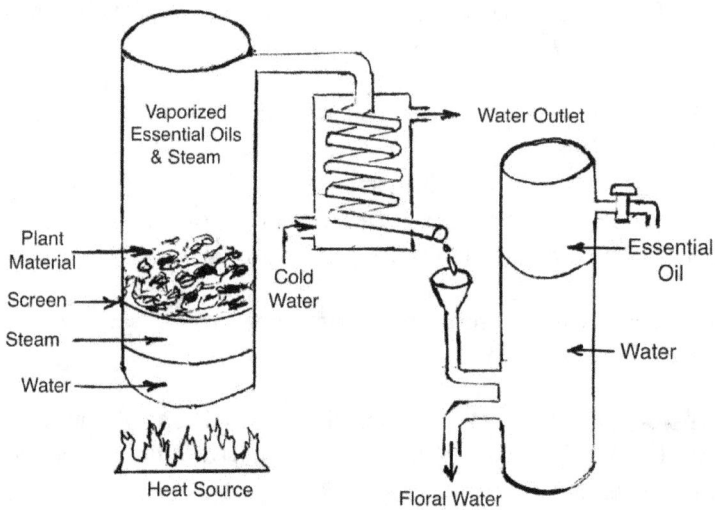

A comparison of yields from laboratory steam distillation of a variety of plant materials is shown in Table 1.2. There is a large disparity of yields, with the flower blossoms providing the lowest yields, while the resins and woody materials deliver higher yields. The re-condensed water, shown as floral water in the figure, is termed a hydrosol which also has commercial value. Commonly available hydrosols include rose water, lavender water, orange blossom water, lemon balm and clary sage waters.

Table 1.2 Laboratory Yields of Essential Oils from Plant Materials

Plant	Botanical Name	Part	Yield, %
Anise	*Pimpinella anisum*	Seeds	0.6-1.5
Camphor	*Cinnamomum camphora*	Leaves & Twigs	2.0
Cassia	*Cinnamomum cassia*	Bark	1.5
Cedarwood*	*Juniperus virginia*	Wood Chips	4.3
Clary Sage	*Salvia sclaria*	Leaves & Tops	0.1-0.3
Eucalyptus	*Eucalyptus sp.*	Leaves & Twigs	1.0-7.0
Frankincense (Olibanum)	*Boswellia carterii*	Gum Resin	3.5-6.0
Geranium	*Pelargonium graveolens*	Leaves	0.3-2.0
Juniperberry	*Juniperus communis*	Berries	1.5
Lavender	*Lavendula sp.*	Flowers	0.5
Orange (Neroli)	*Citrus sp.*	Flowers	0.1
Peppermint	*Mentha Piperita*	Leaves	1.0-2.5
Pine Needles	*Pinus sp.*	Needles	0.5-3.0
Rose	*Rosa sp.*	Flowers	0.006
Sandalwood	*Santalum album*	Wood	4.5

Reference: Yield Guide, Essential Oil Co. (essentialoil.com),
*Not true cedar which is *Cedrus sp.*

Solvent Extraction

A number of essential oils are heat sensitive and become denatured with steam distillation so alternative solvent extraction is utilized, although more expensive. Solvent extraction is also more efficient for plant material containing only very small amounts of the essential oil such as jasmine blossoms.

Solvents such as hexane and petroleum ether are used to obtain the oils which are contaminated with a mixture of waxes, resins and other lipophilic-oil soluble materials. This mixture is termed the concrete and it is necessary to perform another extraction with ethyl alcohol to separate the fragrant oil from the other components. The absolute is then obtained by either a second low temperature distillation to remove the alcohol and leave the absolute; or the waxes and lipids are precipitated out by cooling the ethanol mixture to $0°F$ with the absolute then obtained after filtering and vacuum purging.

A more recent approach for solvent extraction is the use of supercritical liquid carbon dioxide. Under high pressure, CO_2 turns into a liquid which is an inert solvent for extraction of the oil. With release of the pressure, CO_2 simply reverts back to a gas with no solvent residue. This method avoids the loss a very volatile top notes that occurs with steam distillation and also avoids some of the residues from hexane extraction. The initial lower temperature of only 83^0F also inhibits denaturization of the essential oil. The resulting concrete still contains waxes, but these are easily separated by simply lowering the temperature of the pressurized supercritical fluid (6-9, 13-16).

Expression

This method is utilized mainly for citrus peels and rinds. The peels are either pressed mechanically or cold pressed similar to the procedure used for obtaining olive oil. Higher yields are obtained with industrial grinders that force the rinds between spiked rollers that open and release the essential oils. The aqueous emulsion of oil and water is then separated by centrifugation.

With the use of expression, the constituents are not altered by heat, but the oils can be contaminated with other undesirable compounds in the rinds. Depending on the variety of the fruit, the peels have an oil content of 0.5-5.0%. Since there are large quantities of peels and rinds cheaply available and the procedure for pressing is cost effective, citrus fruit oils are the lowest cost on the market. Sweet orange, lemon, bergamot and lime oils are valuable byproducts of the citrus industry.

Tinctures and Resinoids

Tinctures are obtained by macerating plant or animal raw materials directly in alcohol. Although yields are often too low for use in perfumery, tinctures have applications in aromatherapy. Resinoids can be similarly obtained by extraction of plant exudates and resins with alcohol. However, the product is usually viscous, sticky and resinous, therefore essential oils from resins are often obtained by steam distillation or solvent extraction.

Essential Oils and Odorants from Plants

A number of important plant species for obtaining perfume ingredients are discussed in this section (Table 1.1). There are a wide range of plant species and plant components from which essential oils and other odorants can be obtained and range of plant components is clearly demonstrated for the Bitter Orange Tree described below. A more detailed discussion for most of these essential oils and their synthetic substitutes are provided in Chapter 4 on the Chemistry of Odorants. Characteristic perfumes that contain the different essential oil notes in their formulations are also noted in this section.

Oils from the Bitter Orange Tree (*Citrus aurantium*)

An excellent example of a plant which contains a diversity of valuable essential oils in many parts of its structure is the Bitter Orange Tree. The essential oils from this tree were once obtained primarily from Tunisia and Lebanon but now the production is much more widespread including China, India, Brazil and the West Indies.

The essential oil in the blossoms of Bitter Orange is referred to as **Neroli Oil** when obtained by stream distillation and this oil is one of the most valued for use in perfumes. Another "**Orange Blossom Oil**" is obtained by solvent extraction of the flowers and also used in perfumery. Fresh leaves and green twigs of the orange tree are the source of **Petitgrain Oil** which is primarily used in cosmetics and soaps, while the fruit peel (rind) yields the familiar **Orange Oil** (1-5).

Neroli Oil has a very enchanting fragrance so valuable in many perfumes. It is light, floral, citrusy with orange/honey notes. The orange blossoms must be harvested carefully because bruising promotes fermentation and adulteration. Typical yields of Neroli oil are only 1 kg of oil from a ton of blossoms.

Neroli oil is a component of many modern perfumes at concentrations of 10-12% (13). The oil also has therapeutic value as an anti-depressant with profound effects on the limbic system. Primary chemical constituents include linalool, limonene, pinene and linalyl acetate.

Petitgrain Oil has a distinct rosy wood, green floral fragrance. Paraguay is the largest producer of petitgrain oil although it is also produced in France and Morocco. Yields of petitgrain oil are higher, in the range of one liter per 500 lbs. of leaves and twigs, such that the price is more reasonable and the oil provides good performance in colognes, air fresheners and soaps. Because of the high price of neroli oil, it is sometimes combined with petitgrain oil to produce a mix of essential oils referred to as *Petitgrain sur Fleurs* or Petitgrain over Flowers. Primary chemical constituents in petitgrain oil are similar to Neroli oil but in different proportions (8-16).

The third essential oil derived from the Bitter Orange tree is **Orange Oil**. It is obtained by cold pressing the rinds of the fruit and it has a sweet, citrus/floral aroma that lies between mandarin and pomelo. It can be used in both perfumery and aromatherapy. Limonene comprises 95% of chemical composition (8-16). The chemical structures for the constituents of the oils obtained from the bitter orange tree are given in the chapter on Chemistry of Odorants (1,2,8-12).

Some fragrances featuring the various extracts from the bitter orange include *Les Colognes Neroli* by Annick Goutal, a verdant rush primarily composed of neroli, orange blossom and petitgrain oils. Jean-Paul Gaultier launched several orange blossom theme perfumes starting in the 90s as *Classique, Fragile* and *Fleurs du Male* and Tom Ford more recently produced a fresh neroli and orange blossom-based perfume in 2016, *Neroli Portofino* (17).

Essential Oils from Plant Blossoms

A primary source of many essential oils for both perfumery and aromatherapy is from plant blossoms. Over 200 different plant flowers have been utilized in formulations for perfumes, mainly as the heart note, and many are used for aromatherapy. Some of the most common flower fragrances in perfumes are rose, lavender, jasmine, ylang-ylang, tuberose, bitter orange, carnation, frangipani (plumeria), gardenia, honeysuckle, violet, lotus, geranium, iris, lilac, hyacinth, narcissus, orchid and lily-of-the-valley. However, it should be noted that a number of the floral blossoms are either too delicate or contain such minute quantities of the essential oil such that the fragrance is recreated with synthetic ingredients. These florals include lilac, lily, lily-of-the-valley (muget), freesia, honeysuckle, hyacinth, heliotrope, helichrysum, Iris (orris) and violet (1-15).

A few of the most significant flower notes of rose, jasmine, lavender, ylang-ylang, and tuberose utilized in many perfumes are described below. A further description of the chemical constituents and synthetic substitutes for these essential oils, and for a number of other flowers (violet, lilac, geranium, gardenia, carnation, hyacinth, orange blossom, orris root), are given in the chapter on Chemistry of Odorants.

Rose (*Rosa damascena* & *Rosa centifolia*)

Rose is the oldest, most famous, floral fragrance. Of all the flowers, rose is the universal symbol of spirituality in religious culture. There are 5,000 varieties of roses but only two are used for perfume – **Damask Rose** (*Rosa damascena*) and **Rose de Mai** (*Rosa centifolia*). Damask Rose is grown predominantly in the Valley of the Roses in Bulgaria, but also in Syria, Turkey, Russia, India and China. Rose de Mai or cabbage rose is a hybrid grown in France, Morocco and Egypt. The flower is picked in the late bud stage and harvesting of the rose petals can be difficult due to the thorns (8-12).

Rose is considered the "King of Flowers" for its special fragrance. The essential oil or "Otto of Roses" is obtained by steam distillation, while solvent extracted rose oil is referred to as Rose Absolute. The highest quality rose oil with superior intensity is obtained by CO_2 extraction. All the oils are very intoxicating and expensive. The fragrance of rose oil is clean, lemony fresh, floral, dark berry and somewhat boozy liquor-like.

Rose petals secrete only minute amounts of oil compared to lavender, for example, resulting in the very high price of rose oil. One kg of rose essential oil requires four tons of rose petals. The best quality oils sell for as much as $5,000/lb. Geranium oil is often used to adulterate rose oil since it has a similar aroma due to the active ingredient, geraniol.

Rose fragrance is the main ingredient in almost three-quarters of modern prestige perfumes such as *Chanel #5 (Chanel), Joy (Jean Patou), Tresor (Lancome), Paris (Yves St. Laurent), L'Air du Temps (Nina Ricci)*, and *Eternity (Calvin Klein)*. The concentration of rose absolute is 0.5% in *Chanel No.5, Paris (Yves St. Laurent)* and *Angel (Mugler)*, 0.7% in *Perles de Lalique (Lalique)* and 0.12% in *Curious (Britney Spears)* (11-15).

The oil is a potent anti-depressant, brings joy to the heart, promotes feeling of love and relaxes the central nervous system. The main chemical constituents of rose oil are phenylethylalcohol, citronellol, geraniol and nerol.

Jasmine (*Jasminum grandiflorum*)

The highest quality of jasmine is mainly cultivated in France with one supplier owning all the production. Less expensive jasmine absolutes are obtained from Italy, Egypt and Morocco. The yield of essential oil is dependent on the time of harvest over the summer with yields peaking in mid-September for jasmine grown in Liguria, Italy as shown in Figure 1.2. The flowers are very delicate, so to preserve the fragrance, they are hand-picked starting in the late evening, while the buds are collected early in the morning. The jasmine flower is also referred to as the "Queen of the Night" because the flower yields its full aroma in the cooler evenings (8-12).

The name jasmine is of Persian origin meaning gift from God. So often jasmine flowers are utilized in religious and cultural events in many parts of the world. The enchanting aroma symbolizes love, peace, and modesty. In China the blossoms are used to produce a tea and referred to as "Flower Tea."

Jasmine oil is highly esteemed in perfumes and difficult to produce synthetically. The oil is usually obtained by solvent extraction due to the sensitive nature of the blossoms which results in a higher price. Eight thousand fresh flowers yields just a milliliter of the absolute (1000 kg flowers for one gram absolute) (1-10).

After rose, this floral aroma is the most sought-after, and rose and jasmine are the foundations of perfumery. Approximately 83% of women's and 33% of men's fragrances contain jasmine. A perfume without the jasmine is hard to find and the floral smell of jasmine constitutes the heart of many perfumes. The aroma is narcotic sweet floral, exotic, sultry, sometimes greener/airer, with an animalic background. It blends particularly well with citrus oils, bergamot, rose and sandalwood (8-12,16,17).

The highly desired absolute of jasmine can be found at different concentrations in a many notable fragrances such as *Chanel No.5* at 4%, *Bois des Îles (Chanel)* 2.5%, *Arpègé (Lanvin)* 1.5%, *Ma Griffe (Carven)* 0.8%, *L'Heure Bleue* 0.4% *(Guerlain)* and *Cuir de Russie (Chanel)* 0.28% (11-15).

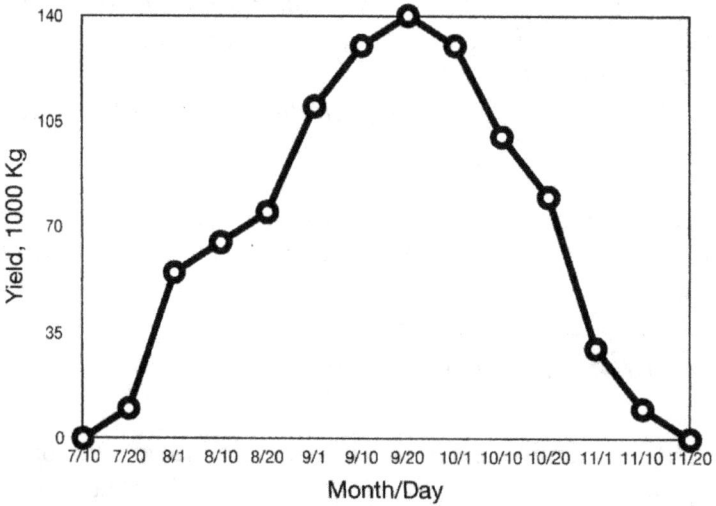

Figure 1.2 Variation in yield of essential oil from jasmine flowers in summer months (Liguria,Italy) (2).

Jasmine fragrance is one of the most sensual scents available to the perfumer and this is purportedly due to the chemical component, indole. In addition to indole, some other principal chemical components present in jasmine oil include benzyl acetate, linalool and cis-jasmone (13-15).

Lavender (*Lavandula augustifolia*)

Lavender is grown extensively in France and England, with France producing 30% of the world's lavender while Bulgaria produces the largest amount of the oil. The flowers are typically harvested in June. The essence has been used in soothing bath waters for thousands of years by the Greeks, Romans, and Persians and referred to as *Lavare* meaning to wash (1-12).

There are 39 species of lavender which produce a variety of scents, but English lavender (*Lavandula augustifolia*) is the one primarily utilized in perfumery. The essential oil is obtained by steam distillation of the flower spikes and the yield is about 1% based on dry weight or 1 kg of essential oil from 20 kg of fresh lavender blossoms (1-10).

The characteristic aroma is aromatic, clean, medicinal, with a licorice-like end. It is closely aligned with Fougère compositions often utilized in men's perfumes. It is also used extensively in soaps, cosmetics, and medicinals. It is purported to have anti-inflammatory properties and a calming sedative effect commonly encompassed in sachets as a sleep aid.

The primary chemical constituents are linalyl acetate and linalool. Lavender is present in many fragrances, such as *Lavender (Avon), Lavender Illusion (Victor & Rolf)* and *Lavender & Coriander (Jo Malone)* (16,17). A further discussion of lavender and lavandin is given in the chapter on Chemistry of Odorants.

Ylang-Ylang (*Cananga odorata*)

Known as the Cananga or Perfume tree, ylang-ylang originated in Asia, notably in the Philippines and Indonesia. However, the main source of the essential oil for many years has been the French Comoros Islands in the Indian Ocean, near the Seychelles and Madagascar. The essential oil is up to 30% of the annual export from the Comoros (8-12,19).

The essential oil is obtained by steam distillation and separated in different grades as the distillates are collected (labelled I, II & III). The intensity of the scent varies from the various oil distillates. Ylang-Ylang extra is the most potent and is often used as a top not in perfumes, notably *Chanel No.5*. The less potent oils are used as middle or base notes in perfumes as well as in colognes, lotions, food flavoring and soap.

Due to the complex nature of the constituents and the aroma profile of ylang-ylang, it is sometimes referred to as the "Flower of Flowers." It is used in very many perfumes as top, middle and base notes and it is thus a very valuable scent in the art of perfumery (8-12,16).

Ylang-Ylang has a delicate evanescent (quickly fading) aroma which is very rich floral and sweet narcotic with exotic tones. It is used in aromatherapy to stabilize emotion, help with depression and relive tension. Ylang-ylang is pervasive in floral fragrances, over 40% of perfumes, two which feature it are *Ylang Vanille (Lys Sloéia)* and *Coeur d'Ylang (Comptoir Sud Pacifique)* (17).

Tuberose (*Agrave amica*)

The plant was first described as *Polianthes tuberosa,* however as an agave plant with a flowering spike, it is now identified as *Agave amica*. Native to southern Mexico it was domesticated by the Aztecs who held it sacred to their goddess Xochiquetzal. It no longer occurs in the wild but is grown world-wide. In India the blossoms are formed into garlands and offered to gods or for wedding celebrations, while in Indonesia tuberose is used in cooking. It is one of the most fragrant white flowers grown in the world (1-5,8-11,19).

Blossoms are picked in bud stage for maximum aroma and, due to the delicate nature of the blossoms, the oil is obtained by solvent extraction. It takes 3600 kilos of flowers to produce a liter of oil. The essence is very strong and it is recommended that it be used in moderation since it can be overpowering. It was a favorite fragrance of French Queen Marie Antionette in the perfume *Parfum de Trianon*. The fragrance is surprisingly carnal, creamy, fleshy floral. Chandler Burr, a curator of olfactory art, referred to it as a "Flower with Claws" while perfumer Roja Dove called tuberose the "Harlot of Perfumery" (8-10,16).

The main chemical constituents are methyl benzoate, benzyl benzoate, methyl anthranilate and pentacosane. Tuberose is an important aroma in many exotic fragrances and contained in over 20% of quality fragrances including *Fracas (Robert Piquet), Carnal Flower (Frederic Malle), Tubéreuse Criminelle (Serge Lutens)* and *Poison (Dior)* (17).

Essential Oils from Leaves

The leaves of many plants are utilized for steam distillation of essential oils. Some of the more commonly utilized plant leaves are basil, bay leaf, bitter orange, buchu, cinnamon, sage, eucalyptus, geranium, guava, lemon grass, manuka, melaleuca, orange, oregano, palmarosa, patchouli, peppermint, rosemary, spearmint, tea tree, thyme, violet and wintergreen. Tree needles are also utilized from cypress, fir, pine and spruce (1-5,8-11). A few significant leaf extracts utilized in perfumes and aromatherapy are described below. A further description of the chemical constituents in patchouli oil is given in the chapter on Chemistry of Odorants.

Lemon Grass (*Cymbopogon flexuosus*)

Lemon grass, sometimes referred to as citronella grass, is widely cultivated in southeast Asia. Two species are utilized for extraction of the essential oil, *Cymbopogon citratus* and *Cymbopogon flexuosus*, with the former used more for culinary applications and the latter for perfumes, soaps, detergents, deodorants, cosmetics and other scented products. It is utilized in aromatherapy for its soothing and calming properties. In contrast, it is also an effective insect repellent commonly used to repel mosquitoes. Because of the many applications and properties of the oil, it is among the top ten most utilized essential oils on the market (1-5,8-12).

The essential oil is obtained by steam distillation which results in greater purity compared to expressed lemon oil. Thus, it is widely used as a substitute for expressed lemon oil. The lemony-ginger sharp odor derives from the dominant chemical component, citral, in the essential oil. Some fragrances with a lemon grass theme include *Green Tea (Ravenscourt Apothecary), Cactus Garden (Louis Vuitton)* and *Verbena (O Boticário)* (17).

Eucalyptus (*Eucalyptus citriodora*)

Eucalyptus is native to Australia but grown world-wide in milder climates. Australian aborigines utilized the leaves for traditional medicine and the earlier explorers promoted the use of the oil as a disinfectant. Simple crushing of the leaves of eucalyptus releases the characteristic clean medicinal scent valuable for both aromatherapy and perfumery.

The essential oil is steam distilled from the leaves and China produces 75% of the world trade of the oil, followed South Africa, Iberia and Australia. There is an extensive list of different Eucalyptus species grown in the world with the greatest production of the oil from *Eucalyptus globulus* (3000 tons/yr.); while for perfumes *Eucalyptus citriodora* is preferred with production at 1500 tons/yr (1-5,8-12).

Eucalyptus oils have widespread applications as pharmaceuticals, antiseptics, flavorings, perfumes and, like citronella oil, it is used as an insect repellent. As a fragrance component the oil imparts a fresh, clean aroma to soaps, detergents, lotions and perfumes. The fragrance is described as fresh, clean, camphor-like with lemony aspects. Boelens et al. (18) described the progression of the aroma of a eucalyptus oil (*E. globulus*) with the initial impression as harsh, terpene-like but moved to a fresh, minty and camphoraceous aroma after one-half hour, followed by a hay, rosemary-like scent after several hours. Finally, a woody and powdery dry-down was noted at over five hours. The chemical components include α-pinene, citronellal, citronellol, eucalyptol, and cineole (13-15). Eucalyptus is featured in a number of fragrances such as *Jangala (Pierre Guillaume Paris)*, *Eucalipto (Granado)* and *Eucalyptus (Thymes)* (17).

Patchouli *(Pogostemon cablin)*

Patchouli is native to tropical Asia where it is grown extensively, notably in India, but it is also grown in South America and the Caribbean. It has been used for centuries in perfumes, alternative medicine and for incense and insect repellents.

An advertisement in a London newspaper in 1846 stated that "Viner's patchouli is confidently recommended as the only remedy known to prevent moths. In foreign countries the peculiar properties of this Indian perfume are highly appreciated, it is therefore most extensively applied to this useful purpose." Dried patchouli leaves were tucked into the folds of exported Indian textiles to deter moths, impregnating them with an unmistakable, musky aroma. In Victorian Britain, the popularity of Indian shawls resulted in the ubiquitous scent becoming a symbol of luxury, as well as distinguishing the material of Indian origin (8-11,19).

The essential oil is recovered by steam distillation of the dried leaves and the leaves may be harvested several times a year. Higher quality oils can be obtained by distilling wet leaves on-site soon after harvest. Another approach more frequently utilized for improved quality is boiling of the dried leaves followed by fermentation (1-5,8-11).

Patchouli is in the lavender family and has been referred to as the "unwashed cousin" of lavender in aroma (16). It is an excellent perfume base note and fixative and occurs in many perfume blends, esp. woody, floral, musk types. The note is rich earthy, woody and moist and is key to the Chypre accord fragrances. Patchouli appears in one-third of the top perfumes. *Miss Dior (Dior)* contains 9.2% patchouli oil. Some othere fragrances featuring patchouli include *Patchouli (Santa Maria Novella), Patchouly Indonesiano (Farmacia SS Annunziata)* and *Psychedelique (Jovoy)* (17).

Patchouli is also important to Gourmand notes where it counter-balances the sweet sugary scents of crème, chocolate and caramel in fragrances such as *Angel* and *A*Men* by Thierry Mugler. Two important chemical components of the essential oil are patchoulol and norpatchoulenol.

Essential Oils from Roots & Rhizomes

The rhizomes and roots of a number of plants have been utilized for collection of essential oils. Some of the more commonly utilized plants in this class of essential oils are angelica, costus, ginger, orris (sweet iris), sassafras, spikenard, valerian and vetiver (1-5,8-11). A few significant roots and rhizomes utilized in perfumes and aromatherapy are described below and a further description of the chemical constituents in vetiver and orris root (sweet iris) essential oil are given in the chapter on Chemistry of Odorants.

Vetiver (*Chrysopogon zizanioides*)

Vetiver is a perennial bunchgrass grown in northern and western India where it is known as "khus." It is also grown in Indonesia and the Seychelles, however 50% of the world's supply is from Haiti. The roots may extend as deep as 12-feet into the ground and the tufts of stems grow up to six feet tall with thin, rigid leaves and purple-brown flowers (1-5,8-11).

The essential oil is steam distilled from the roots and utilized in perfumes, cosmetics, herbal skin care and aromatherapy. The antiseptic properties are useful for treating acne and sores. Worldwide production of the oil is 300 tons/yr (1-5,8-19).

Vetiver is used widely in perfumes for its excellent fixative properties. It is contained in 90% of western perfumes and is a more common ingredient in fragrances for men. The fragrance is deep, sweet, woody, smoky, earthy, amber, and balsam. *Mitsouko (Guerlain)* and *Vetiver Pour Homme (Carven)* contain high contents of vetiver oil, 10% and 9.3% respectively, while *Femme (Rochas)* has 3.5% and *Arpège (Lanvin)* only 0.65%, but supplemented with 11% of the synthetic, vertiveryl acetate (13-15).

Some other vetiver featured fragrances include *Vetiver (Guerlain), Vetiver Extraordinaire (Frederic Malle)* and *Sycomore (Chanel)* targeted to women (17). Like all essential oils, vetiver is a complex mixture, with some unique chemical structures such as alpha-vetivone, khusimol along with more common terpinenol and benzoic acid.

Costus (*Saussurea costus*)

The oil is obtained by steam distillation of the dried root of the *Saussurea costus* plant. The oil has a unique "wet dog" aroma which provides a special nuance to Oriental type perfumes. The odor has been described variously as soft, woody, tenacious, similar to vetiver and even fatty. Often utilized in unisex fragrances such as *Notes (Robert Piquet)* and *Othoca Sandalia* (Sandalia). The boldest use of this rare fragrance is in a more recent launch in 2019 of *Black Sheep* by the House of Matriarch (17).

Essential Oils from Seeds

Seeds utilized for collection of essential oils are numerous and include both familiar and unusual species. The most significant are almond (bitter), ambrette (hibiscus), anise, aniseed, black pepper (peppercorns), cacao, caraway, cardamom, carrot, celery, coffee bean, coriander, cumin, dill, fennel, nutmeg (mace), tonka bean and vanilla (fermented) (1-5,8-11). A couple significant seeds utilized to obtain extracts for use in perfumes are described below.

Tonka Bean (*Dipteryx odorata*)

Tonka beans are obtained from the seeds of the Cumaru tree (*Dipteryx odorata*) which can grow up to 90 ft tall and is native to Central and South America. The name "tonka" is from the Carib language spoken by natives of French Guiana. The main producers of tonka beans today are Venezuela and Nigeria (1-5, 8-11).

Strictly tonka bean oil is not an essential oil since the oil is not obtained by steam distillation. The seeds contain up to 46% oil based on dry weight. Coumarin is responsible for the pleasant odor of the seed and the natural coumarin is obtained from the seeds by soaking in alcohol for 24 hours, after which a white powder accumulates on the surface of the seed. This powder is composed of crystals of coumarin. However, now most commercially produced coumarin is synthetic, which has reduced the demand for tonka beans as a crop (9-11,19).

Coumarin is used in the perfume industry and as a flavoring, particularly for tobacco. It has a bitter taste and is toxic at very high concentrations. Like a number of other plants, the tonka bean plant probably produces coumarin as a defense chemical. However, in France, tonka beans are used in in desserts and stews (9-11).

The sweet, warm fragrance of coumarin has been likened to summer meadows, the fragrance of new mown hay with a bittersweet almond component. It is gently uplifting and light and has a versatile tone useful for blending into perfumes. The aroma is notable in fragrance such as *Tonka Imperiale (Guerlain), Tonkade (Laboratorio Olfacttivo)* and *Tonka (Reminiscence)* (17).

Vanilla (*Vanilla planifolia*)

Vanilla is native to Mexico and Central America. The flower has only a limited number of specific pollinators so almost all vanilla grown in the world is hand-pollinated. After planting, it takes four years for blossoms to appear and only one kilo of pods is produced from 600 pollinated blossoms. The pods are picked while still green, dried and the fermented for extraction of the essential oil. Both steam distillation and CO_2 solvent extraction are used to recover the oil (1-5,8-12).

The availability of natural vanilla has been severely depleted due to the huge global demand, a series of poor harvests and the effects of typhoons. As a result, the cost has increased dramatically. The primary chemical components are vanillin, ethyl vanillin and coumarin and all of these chemicals can be produced synthetically. However, they do not necessarily provide the rich deep tones of natural vanilla extracts nor do they adequately provide the therapeutic benefits (8-12).

The fragrance of vanilla has universal appeal and it is used in very many perfume formulations where it provides a deep, sweet, rich aroma with caramel and chocolate nuances and deep balsamic undertones. Vanilla is used to soothe and calm the psyche and help with frigidity in aromatherapy. Some fragrances featuring a vanilla aroma include *Vanille (Molinard), Spiritueuse Double Vanille (Guerlain), Vanille Orchidée (Van Cleef & Arpels)* and *Vaniglia del Madagascar (I Profumi di Firenze)* (17).

Cardamom (*Elettaria cardamomum*)

Cardamom is a member of the ginger family and has been used in food, medicine and perfumery for many years. It is commonly added to baked goods in Sweden where the shops acquire a cardamom aroma atmosphere. Cardamom is grown in Guatemala, Honduras and Sri Lanka and is known to some as an aphrodisiac (1-5,8-

12).

The seed is typically steam distilled to recover the essential oil, although CO_2 solvent extraction is also utilized for a slightly different product. The fragrance is uniquely spicy-sweet, very aromatic and penetrating with balsamic-woody overtones. The tenacious aroma is valuable as a middle note in perfumes (8-12).

Cardamom also has a long history of use in alternative medicine. For Ayurveda it is said to "kindle the fires of digestion, stimulate the activity of the heart and refresh the mind." The oil has antibacterial and anti-inflammatory properties and is also useful for treatment of respiratory problems. The major chemicals in the oil are terpinyl acetate, 1,8-cineole, linalyl acetate and sabinene (13-15). Some fragrances featuring cardamom are *Cardamom (Demeter), Cardamusc (Hermès)* and *Cardamom Coffee (Gorilla)* (17).

Essential Oils from Fruit

This category includes fruits, berries and citrus rinds. Citrus notes are prolific in perfumes and are easily obtained directly from the rinds of the fruit or they can be produced synthetically. Fruit rinds typically utilized for essential oil production by expression (pressing) include bergamot, grapefruit, lemon, lime, mandarin, orange and tangerine. Table 1.3 shows the plethora of citrus notes in a few significant perfumes (13). Bergamot oil is very valuable in both perfumery and aromatherapy. It is often used as a top note in perfumes and is commonly used in aromatherapy to elevate mood and alleviate stress.

Fruit aromas have become more common in poplar modern perfumes and include apple, apricot, cherry, coconut, fig, grape, mango, peach, pineapple and strawberry (8-12). It is important to note that it is almost impossible to extract natural fragrance from fruits and vegetables by distillation, so synthetic chemical fragrances are generally used to recreate the aroma for perfume applications. Some examples of synthetics used for fruit odors are given in the chapter on Chemistry of Odorants.

Berries and buds are also used for collection of essential oils and include berries from juniper, *Litsea cubeba,* and star anise and plant buds from black currant (cassis) and birch trees. A few examples are given below.

Table 1.3 Citrus Oil Notes in Some Significant Fragrances

Perfume	House	Bergamot,%	Lemon,%	Orange,%	Mandarin,%	Lime,%
Eau Savage	Dior	35	20	2	-	-
CK One	Clavin Klein	8	1	2	1.4	0.5
Allegoria Pamplelune	Guerlain	-	20	14	-	-
Jicky	Guerlain	32	2	-	-	-
Coco	Chanel	1.1	5.2	0.3	1.9	-
Cristalle	Chanel	8.3	8.3	-	-	-

Reference: Ohloff et al. (13)

Black Currant (*Ribes nigrum*)

Black currant is originally from the British Isles and is now cultivated in Northern Europe, especially in France. The black currant buds are harvested from December to February and solvent extracted to yield about one pound of absolute per 300 pounds of buds. Buchu leaf oil has also been used to provide a subtle hint of black currant since it displays certain aspects of the odor (8-12,19).

The black currant fragrance is intense, providing a rare fruity top note for perfumes. The dominant tangy green-fruity aroma combines with minty-citrus notes and warm earthy undertones to produce a special fragrance. There is also a hint of a unique animalic allure which can be traced to minute amounts of the sulfur compound, 4-methoxy-2-methylbutan-2-thiol (13-15). Black currant is featured in fragrances such as *Blackcurrant Angel (Lush), Flowerbomb Midnight (Viktor & Rolf)* and *Far Away Glamour (Avon)* (17).

Juniper Berry (*Juniperus communis*)

The berries are harvested from the Juniper trees in Macedonia, Bosnia and Nepal. They are either steam distilled or CO_2 solvent extracted depending on the applications. The CO_2 solvent extracted oil more closely resembles the aroma that occurs naturally from the tree, but both oils have a recognizable aroma that reminds some people of gin, which of course, is flavored with juniper. The fragrance is a radiant woody-fruity aroma with green, somewhat sweet undertones (8-12).

The oil is also valuable in aromatherapy for alleviating mental exhaustion and for asthma, hay fever and nervous tension by inhalation. The chemical constituents include α-pinene, sabinene and terpineol. Juniper berry aroma is noticeable in perfumes such as *Original Santal (Creed), Oligarch (Roja Dove)* and *Date for Men (Fragance One)* (17).

Essential Oils from Wood and Bark

Wood from several different trees is a source of valuable essential oils. Commonly utilized wood includes agarwood, amyris, cedarwood, rosewood and sandalwood. The bark of several trees is also important for collection of essential oils including cassia and cinnamon (1,2,8-12). A few of the most important woods for essential oils are described below and a further description of the chemical constituents in sandalwood is given in Chapter 4 on Chemistry of Odorants.

Sandalwood (*Santalum album*)

The sandalwood tree is a medium size evergreen found in India, Australia and formerly in Hawaii. It was wiped out from Hawaii when cut down by the Hawaiian monarchy to pay off a huge debt to China for imported porcelain. Also, the Indian government has banned the export of sandalwood and sandalwood oil due to their dwindling resources. There are now sandalwood plantations in Indonesia and Australia (1-3,8-16).

The sandalwood name is derived from Sanskrit meaning "wood for burning incense" and the Latin *candere*, to shine or glow. Indeed, the wood alone has a strong, long-lasting fragrance and has been used for incense for 4,000 years. Artist wood carvings made from sandalwood can hold their fragrance for years (8-12,19).

The tree must be destroyed for collection of the oil. The sandalwood oil is typically obtained by steam distillation of the ground wood and the yield varies with the age and location of the tree. Older trees provide the highest oil content and quality. The tree should be at minimum of 15 years old and 30 years old for optimum production of the sandalwood oil (1,2,8-12).

Sandalwood oil is in high demand due to the very pleasant aroma and use as an excellent fixative in perfumes. It is also used in cosmetics, soaps, flavoring and aromatherapy. Due to the need to remove the trees for oil production, the wood is one of the most expensive in the world. The price can increase 10x in 10 years, and as a result there has been widespread smuggling of the wood.

The enchanting fragrance is woody, milky, soft, sturdy, rich, with a green top note and a most satisfying lingering scent. The fragrance *Samsara (Guerlain)* has a very high dosage of sandalwood oil of 20% and *Bois des Îles (Chanel)* with 15.8%. *Sandalwood Temple (Sana Jardin)* is based on ethically sourced East Indian sandalwood while *Santal 33 (Le Labor)* utilizes Australian sandalwood which is less dense and buttery than other sandalwoods (13-15). However many modern fragrances now use synthetic analogs due to cost considerations. Some other fragrances with a sandalwood theme include *Sandalo Nobile (Nobile 1942), Santal Blush (Tom Ford)* and *Sandalwood Attar (Ajmal)* (17).

About 75% of sandalwood oil is the chemical, santalol, and considerable efforts have been made to produce a variety of synthetic sandalwood analogs, further described in the chapter on Chemistry of Odorants.

Rosewood (*Aniba rosaeodora*)

Rosewood essential oil is sometimes referred to as "Bois de Rose" essential oil. It is obtained from the rosewood tree that, like sandalwood, must be destroyed to obtain the oil and as a result it is now an endangered species (1,2,8-12).

The oil is recovered by steam distillation of rosewood chips. The fragrance is subtle and surprisingly sweet, with woody and fruity notes and a floral quality. Rosewood oil is versatile for use in perfumes and blends well with other wood, citrus, spice, herbaceous and floral oils. It has been found to add a special spicy vibrance to lily-of-the-valley type perfumes. It also has a wide variety of applications for aromatherapy (8-12).

Some major chemical constituents are linalool, terpineol and linalool oxide. The percentage of linalool content of the oil ranges from 73-99% in various woody parts of the tree. A number of perfumes are named rosewood, but do not necessarily contain the rosewood note, some which do are *Amber & Rosewood (Label), Rosewood & Sandalwood, Cedar (Zielinski & Rozen)* and *Rosewood (Illuminum)* (17).

Cedarwood (*Cedrus atlantica*)

In biblical times cedar essential oil was obtained from Cedar of Lebanon, *Cedrus libani,* but in more recent years it is obtained from Atlas Cedar, *Cedrus atlantica.* Both cedar species are considered "True Cedar." Some Juniper species are also referred to as "cedar" but they are a different Genus. True cedar originated in Morocco and Algeria but is now grown around the world, notably in the Mediterranean region; however, Morocco is still a primary source of the essential oil (1,2,8-12).

The oil is steam distilled from the ground wood with yields of 3-5%; the leaves are also used to produce an essential oil with different properties. The fragrance of the wood derived essential oil is floral, fruity and woody with back notes of honey and spice. This compares with the essential oil from the non-true cedar, Virginia Cedar-*Juniperus virginiana,* which is significantly different as a smoky, woody, coniferous, fruity aroma (1-5,8-12).

The iris feminine fragrance *Hiris (Hermès)* contains 2.8% Texas cedarwood oil and 1.4% Viginia cedarwood oil. Some other cedarwood inspired fragrances include

Féminité du Bois (Shiseido), L'Enfant Terrible (Jovoy) and *Cèdre-Iris (Affinesence)* which contains a blend of Virginia, Texan and Atlas cedars. The essential oil of true cedar contains the chemicals cadinene, atlantone and cedrol (8-15).

Essential Oils from Plant Exudates (Resins)

Tree exudates as resins are also important to both perfumery and aromatherapy and have been utilized for centuries (Figure 1.3). The most common are benzoin, balsam of Peru, camphor, elemi, frankincense, galbanum, labdanum, myrrh and opopanax. A few of the most important resin extracts for essential oils are described below and a further description of the chemical constituents in frankincense, labdanum and galbanum are given in Chapter 4 on Chemistry of Odorants.

Frankincense (*Boswellia sp.*)

There are five different species of *Boswellia* utilized for production of frankincense essential oil and the trees are grown in Somalia, Ethiopia and India (Figure 1.3). Frankincense, also known as **Olibanum**, has been an item of commerce for 6,000 years, first as an incense from Arabia and later as an essential oil (1,2,8-12).

Figure 1.3 Frankincense trees (inset - native scarifying bark for collection of resin for essential oil extraction).

Frankincense is symbolic of holiness and righteousness and has been burned as incense as an offering to God. It is thought to ward off evil and disease. Indeed, it was "The Gifts of the Magi" of Gold, Frankincense and Myrrh that were provided by the three Kings as an offering to the baby Jesus in Christianity.

The process for obtaining the essential oil involves wounding the tree by scarifying the bark which results in resin oozing from the wound. The resin droplets are collected, dried and ground for subsequent steam distillation. Frankincense oil is a complex bouquet of 200 chemicals, the concentration and ratio of which are affected by species, season, elevation and microclimate (8-12).

Frankincense oil is highly desired for both perfumes and aromatherapy. It has a surprisingly beneficial mood effect when utilized in aromatherapy and is a valuable base note in perfumery. The fragrance is sweet, woody, balsamic, piney and citrusy (8-12). The chemical composition includes alpha- and beta-pinene, limonene and alpha-thujene but it does not include boswellic acids which have insufficient volatility to vaporize in steam distillation (also see chapter on Chemistry of Odorants). A few fragrances featuring frankincense include *Coromandel Parfum (Chanel), Mandala (Masque Milano)* and *Frankincense & Myrrh (Kuumba Made)* (17).

Balsam (*Myroxylon balsamum*)

The primary balsam essential oil is often referred to as Balsam of Peru. This name was acquired from where the Spaniards shipped the raw product to Europe in the 18th century. It is now mainly produced in El Salvador from the Myroxylon species of trees (1,2,8-12).

The process of producing Balsam of Peru from the Myroxylon tree differs from the procedure for other resins. Strips of bark are removed from the tree and the exposed cambium secretes the balsam resin. The resin is collected by soaking into rags wrapped around the tree which are then boiled in water. The naturally heavier balsam sinks to the bottom and the water on top is discarded. One tree will yield about 1 kg of balsam year (8-12).

Balsam of Peru is used for flavoring of foods and drinks, in medicine and pharmaceutical items for healing properties and in perfumes and toiletries. The fragrance has a somewhat sweet scent of cinnamon and vanilla with a soft green olive base note. The oil contains cinnamic acid, benzoic acid and about 40% is composed of the benzyl and related esters of these free acids. Fragrances with a Peru balsam base note include *Sunrise in Cadaquès (Salvidor Dali), Elixir des Merveilles Ltd Ed. Collector (Hermès)* and *Semi-Bespoke No. 5* (16,17).

A different, but similar, essential oil is collected from a different species of Myroxylon (*Myroxylon toluiferum*) with the name Tolu Oil. It has similar applications to that of Balsam of Peru but has a different fragrance profile and is collected by a different procedure. The aromatic, viscous resin is obtained by carving V-shaped wounds in the bark of *Myroxylon balsamum* trees similar to the procedure for collection of the resins from other trees in this category. The hardened beads formed on the trunk are collected and steam distilled to produce the essential oil (1,2,8-12).

Both oils have a soft tone but, Tolu oil is fresher and sweeter compared with the fragrance of Balsam of Peru which is somewhat more earthier and bitter (8-12). The composition of Tolu oil is similar to that of Balsam of Peru but in different proportions. Fragrances featuring tolu oil include *Ambre (Signature), Prada L Femme Absolu (Prada)* and *Tolu (Ormonde Jayne)* (17).

Benzoin (*Styrax benzoin*)

Benzoin is also commonly referred to as Styrax and is derived from exudates of scarified bark of the Styrax benzoin tree. The trees are commonly found in Sumatra in Indonesia where the majority of the resin is produced (1,2).

After the styrax benzoin tree trunk is cut, the white styrax sap appears, but is not harvested until the styrax sap hardens on the tree trunks. The hardened sap is then ground and extracted with alcohol. Uses for the oil are as a tincture in aromatherapy, for skin care, as a food flavoring, for cosmetics and in perfumery. The fragrance is sweet, balsamic, woody, fruity and floral and it is utilized in perfumes both for aroma and as a fixative (8-12).

Styrax benzoin contains cinnamic acid, benzoic acid, benzaldehyde and vanillin. As with other resins derived from trees, the chemical composition varies with geographical locations and climatic conditions. Fragrances with benzoin in the formulation include *Shalimar Souffle Intense (Guerlain), Moustache Eau de Parfum (Rochas)* and *Cashmere (Franck Boclet)* (17).

It is important to note that a very similar name, Storax, is used for the oil extract from steam distilled resin of sweetgum trees, *Liquidambar sp*. This is a distinctly different product that is sometimes utilized as a fixative in perfumes and for medicinal purposes (8-12).

Labdanum (*Cistus ladaniferus*)

Labdanum is a resin contained in leaves and twigs of Gum Rockrose (*Cistus ladaniferus*). In ancient times it was used as an incense and to treat colds and rheumatism. The main source of the resin is from Cyprus, Crete, France and Spain (1,2).

The raw gum is dark brown to black and is produced from the leaves and twigs by solvent extraction for the absolute or steam distillation of whole branches for the essential oil. Labdanum is highly valued and used extensively, appearing in one-third of modern perfumes, due to its woody-amber and leathery odor and resemblance to Ambergris, a whale extract described below (8-12,19). It is found in dosages of 0.4% in *Pour Monsieur (Chanel)* and *Mitsouko (Guerlain)*, 0.3% in *White Linen (Lauder)* and 0.15% in leathery *Duro (Nasomatto)* (13,17).

Galbanum (*Ferula galbanifkua*)

Reference to galbanum can be found in the ancient writings of Hippocrates and in Pliny's Natural History. It was used for a variety of medicinal uses and is now utilized in aromatherapy as well as perfumes. Galbanum has excellent fixative properties for perfumes and provides a rich, spicy green scent to fragrances. It can be used as a top green note or as a base note in combination with musk or oakmoss especially for Chypre and Fougère type perfumes (8-12,19). The essential oil is widely utilized in perfumery usually in low dosages of less than 0.1%, however the perfume, *Vent Vert (Pierre Balmain)*, has a high dosage of 8% and it is a prominent note in *Chanel No.19*. Galbanum also provides a green note to formulations of *Viper Green (Ex Nihilo), Sport Scent (Jovan), L'Ombre dans l'Eau (Diptyque)* and *Elements Aqua (Hugo Boss)* (13,17).

Essential Oils from Other Plant Sources

There are a few other important sources for essential oils and odorants but oakmoss has traditionally been one of the most important. However, the use of oakmoss in perfumes has been dramatically reduced in recent years due to health hazards.

Oakmoss (*Everenia prunastri*)

The lichen *Everenia prunastri* is a valuable source of the essential oil, oakmoss, which has a deep, earthy aroma and has been extensively utilized as both a base note and fixative in perfumes, particularly for Chypre and Fougère accord perfumes. It is harvested in Macedonia and France and extracted with benzene or hexane, then ethanol to yield the absolute. The odor of oakmoss is mossy-woody with characteristic phenolic and smoky, almost animal-like nuances (8-12,19).

However, the use has been severely restricted (0.1% in perfumes) by the 43rd IFRA amendment as a potential skin irritant which has had a significant effect on perfume formulations. Two chemicals in the oil, atranol and chloroatranol, have been implicated as the culprits. Removal of these chemical constituents from the oakmoss extracts adds significantly to the cost such that there has been a significant decline in consumption of oakmoss (13). Both *Mitsouko (Guerlain)* and *Ma Griff (Carven)* have been carefully reformulated with good results to comply with the ruling. A further description of the chemical constituents in oakmoss and synthetic substitutes are given in Chapter 4 on Chemistry of Odorants.

Odorants from Animal Sources

Animal Odorants have been obtained from a variety of species and have very unique aromas. These scents have been highly valued since they produce a remarkable effect on the other ingredients of a perfume. Many of the odorants obtained from animals have excellent fixative properties and serve as superb base notes creating a balanced perfume. However, most of these sources have been substituted with synthetic chemicals due to the inhumane practices for collection of the oils from the animals. Many of the odorants cannot be obtained without killing the animal. Work is in progress to develop sustainable and humane methods for obtaining the odorants from some of the animals, but most of the aroma chemicals are now chemically synthesized to simulate the animalic scent. A brief introduction is given for each of the animal scents to provide a prospective on their use in classical perfumes and the scent that synthetics attempt to simulate (8,13,19,32,36,103,104).

Sperm Whale (Ambergis)

Ambergis is a waxy, solid substance with a grey-blackish color produced in the digestive system of sperm wales. It has a fecal, marine type odor but as it ages it acquires a sweet, earthy, animalic fragrance. The whale is found primarily in the Atlantic Ocean and collected commercially in Bermuda (1,2).

Ambergris can be found floating on the sea or washed up on coastlines but it is rare and may float for years before landing on the shoreline. The beaks of giant squids have been found in lumps of the ambergris indicating it is produced by the whale to ease the passage out of the gastrointestinal tract.

The essential oil is obtained from the waxy lumps by grinding and extracting with alcohol. The primary use for ambergris is as a fixture and fragrance much like musk in perfumes and perfumes can still be found with ambergris. However, due to the rarity of the substance it has mostly be replaced with synthetic ambroxide.

The fragrance is sweet, dry and musky with stronger notes of wood, moss, and amber and a slight animalic component. The ambergris lumps are composed of oxidized fatty compounds of ambrein, ambroxan and ambrinol.

American Beaver (Castoreum)

The odorant castoreum is obtained from the odorous sacs in the anal area of the North American beaver. The sacs are dried to provide a thick paste that is solvent extracted to yield the oil. Unfortunately, the beaver must be killed to harvest the dried sacs (103,104). Castoreum has been used since ancient times as a medicine to boost the immune system and it has anti-inflammatory and antipyretic effects due to the presence of salicylic acid and its derivatives.

However, castoreum has been mainly produced for perfumery. It has a strong animalic, warm, sweet smell, with leathery nuances mainly used in leather, animalic and Chypre perfumes. It contributes an erotic, spicy, amber note to leather type perfumes, as well as being a fixative.

The oil contains at least sixty compounds with phenol playing and important role. Additional chemicals found in the oil include acetophenone, benzyl alcohol, benzoic acid and traces of ammonia, all contributing to the unique aroma

Civet (Civet Musk)

The civet is a cat-like animal with a long tail and pointed muzzle, related to the mongoose. The musk has been obtained from the odorous sacs of African and Indian civets, with most of the commercial musk collected from caged animals in Africa (103,104).

The animal is sacrificed to obtain a paste from the perineal glands and the paste is solvent extracted to produce an absolute. At full strength the musk smells fecal and nauseating, but when diluted it has a radiant, velvety, floral scent very valuable to the perfumer. The musk adds shimmer, diffusion and warmth to a perfume as well as acting as a fixative. The odor of civet oil is derived from civetone and a variety of ketones, with smaller amounts of indole and skatole.

The high cost and inhumane treatment of the animals provoked a concerted effort to procure synthetic musk substitutes. A number of synthetic musk molecules are now available which are discussed further in Chapter 4 on the Chemistry of Odorants.

Asian Musk Deer (Deer Musk)

Musk deer live mainly in forested and alpine scrub habitats in the Himalaya mountains of southern Asia. There are seven different species of musk dear but they are not a true deer since they exhibit some different physical characteristics (103,104).

The oil is obtained from a hairy pouch the size of a golf ball from the genital area and, like the beaver, the deer must be sacrificed to obtain the musk pod. The pod is then chopped and soaked in alcohol to obtain the oil.

Natural musk was used extensively in perfumery until the late 19th century when economic and ethical motives led to the adoption of synthetic musk which is now used almost exclusively. The fragrance has a sweet, animalistic, earthy-woody aroma with an aromatic intensity and longevity prized as an excellent fixative. It provides both elegance and radiance to a perfume. The compound primarily responsible for the characteristic odor of musk is muscone.

Rock Hyrax (Hyraceum)

The rock hyrax is a medium-sized terrestrial mammal native to Africa and the Middle East. It is found at elevations up to 14,000 feet living in rock crevices to escape predators. The hyraceum oil is obtained from the petrified rock-like excrement of the animal, so it does not involve sacrifice or pain as is the case for oil retrieval from civet, beaver and musk deer.

The hyrax rocks are ground and ethanol solvent extracted to yield an oil with an intense, complex, heavy fermented-type scent that is a cross between the scent of civet and castoreum, with aromas of tobacco and agarwood (103,104). The fossilized hyraceum has also been utilized in traditional South African folk medicine for treatment of epilepsy.

Honey Bees (Honeycomb)

Honeycombs from bees are solvent extracted to produce a warm, sweet oil. The absolute has a stronger aroma than honey and is more musky and animalic. This is due to the unusually large amount of naturally occurring phenylacetic acid. The aroma of this acid when isolated is also described as honey sweet and animalic, with malt-vanilla undertones of extraordinary tenacity. The oil is an excellent fixative in perfumes and can be used in place of natural animal musks (9-12).

Synthetics

Synthetic substitutes for essential oils are now pervasive on the market and it is necessary to check carefully if a natural derived essential oil is the desired product. This is especially important for aromatherapy applications. It is also important to note that all essential oils are composed of chemical compounds and the natural oils can have hundreds of different molecular structures, so it is often difficult to duplicate the exact aroma and nuances of the natural oil with the synthetics. However, specific chemical structures which are the predominate factor affecting the aroma can be isolated and/or synthesized to create the effect of the natural oil (13-15,20-25).

Many new perfumes are primarily composed of synthetic chemicals. The synthetics were first used in perfumery in 1882 and are very important to modern perfumery. Synthetics provide a special sparkle and radiance to perfumes difficult to achieve with the natural oils alone. They can also provide new odors not previously available to the perfumer. Several approaches for obtaining synthetics are described below.

Isolates

Isolates are the individual chemicals which make up a natural essential oil. Jasmine is composed of 900 different chemical structures but only 300 have been isolated for further applications. Many of the chemicals in the oil are in trace amounts which are difficult and/or too expensive to seek out. However, as perfumer Roja Dove has pointed out, sometimes a new isolate is discovered that has a significant effect on perfumery (16). When methyl dihydrojasmonate was isolated from natural jasmine essential oil it was found to have a long-lasting citrus aroma, not at all typical of the common citrus smelling perfume components. This fresh, long-lasting floral citrus fragrance was instrumental to the considerable success of Dior's *Eau Sauvage* perfume created by Edmond Roudnitska. Methyl dihydrojasmonate is extensively utilized in a wide variety of perfume compositions. A list of some important plant chemical isolates is shown in Table 1.4 (20).

As chemical science has evolved over the many years, more sophisticated tracing and isolation methods have been realized making it possible to explore more alternate synthetic compounds for perfumery. The advent of Gas Chromatography-Mass Spectrometry (GC-MS) was a boon to perfumery since the methodology allowed the perfumer to both analyze and isolate chemical components of essential oils as well as reveal the composition of both historical perfumes and those on the market (13-16). GC-MS is further discussed in Chapter 4 on the Chemistry of Odorants.

Once a chemical is isolated from the natural oil and the structure elucidated, it can be synthesized in greater quantities for use in perfumery at a reduced price. However, as previously pointed out the isolates do not possess the subtlety of the whole oil with its myriad of components.

Table 1.4 Chemical Isolates from Plants

Plant Source	Chemical Isolate
Basil (sweet)	Methyl cinnamate
Bitter almond	Benzaldehyde
Caraway	D-Carvone
Cinnamon bark	Cinnamaldehyde
Citronella	Geraniol, citronellol, citronellal, hydroxycitronellal
Citrus	Limonene
Clove	Eugenol
Coriander	Linalool
Grapefruit	Nootkatone
Jasmine oil	Methyl dihydrojasmonate
Lemon	γ-Terpinene, citral
Lemongrass	Geranyl acetate, citral
Lime	α-Terpineol
Melaleuca viridiflora	Nerolidol
Mandarin petitgrain	Dimethyl anthranilate
Mint	*cis*-3-Hexenol
Mint (bergamot)	Linalyl acetate
Orange	Acetaldehyde, decanal, ethyl butyrate, myrcene, nonanal, sinensal, valencene
Orris	α-Irone
Rose geranium	Phenyl ethylalcohol
Spearmint	L-Carvone
Styrax	Cinnamyl cinnamate
Vanilla	Vanillin
Vetiver	Vetiverol

Reference: Rowe (20)

Synthetic Essential Oil Reconstructions

When a natural oil is very rare and/or extremely expensive, the perfumer can "recreate" the natural oil utilizing both less expensive synthetics as well as some natural components. However, these reconstructions can vary considerably in quality. This approach was utilized with synthetic civetone to recreate the civet musk that is in very limited supply and expensive. It is also almost impossible to obtain a suitable amount of natural oil from a number of blossoms including lilac, honeysuckle, freesia and lily-of-the-valley, etc. so the reconstructed oil is the only alternative (16).

Pure Synthetics

The pure synthetic category includes synthesis of chemicals identified and isolated from the natural oils and new discoveries of chemicals with unique and sometimes startling aroma profiles (13-15,21-25).

New aroma chemicals are also discovered when working out the synthesis of an expensive or rare aroma molecule. This is exemplified by the work of the French chemist, Jacques Valliant, working to produce an inexpensive route to a molecule with the scent of sandalwood in the 1970s. Over one year, he synthesized forty-five different analogues of beta-santalol but only one smelled of sandalwood, the rest provided a plethora of scents including peach, cedar wood, lemon grass, rosewood, camphor and cut grass! (26).

Categories of Synthetic Perfume Ingredients

Synthetic chemicals important to perfumery and aromatherapy are differentiated in a number of ways. The categories are based on chemical structure and a common chemical unit (functional group) that is part of the chemical structure of the molecule (13-16). Some of the important categories are shown below with a few examples for each, a more complete listing and description, with chemical structures, are given in the chapter on Chemistry of Odorants. Appendix I provides a list of some of the many important synthetic compounds used in perfumery.

Aldehydes and Ketones

This group includes many chemicals of great importance to modern perfumery. The category includes aliphatic straight carbon chain structures as well as more complex structures containing the characteristic carbonyl functional group.

The dramatic success of perfume *Chanel No.5* is often attributed to the inclusion of a mixture of the aliphatic straight carbon chain aldehydes in the formulation, adding

sparkle and freshness to the perfume. Some of the complex structures in this category are:
>   **Damascones** provide long-lasting fruity rose notes in perfumes;
>   **Ionones** replicate the aroma of violet and were a major discovery in 1893;
>   **Calone,** also known as watermelon ketone, provides the aroma of the seashore and sea breezes with concomitant marine and ozone nuances.

Alcohols

This category includes the chemical most of us know, ethanol the intoxicating liquid, but also valuable for dilution of perfume products and for extraction of essential oils. All the compounds in this class contain the hydroxyl functional group. Some notable complex alcohol structures valuable to the perfumer in this classification are:
>   **Geraniol**, an isolate from rose, palmarosa and geranium is synthesized for its fresh, sweet, rose-like scent;
>   **Ethyl Maltol** or Veltol is both an aldehydic and alcoholic synthetic compound which has an intensely sweet, cotton candy aroma with carmel and chocolate notes.

Esters

Esters are formed from a combination of an alcohol and an acid and are known for their intense fruity aromas. Figure 1.4 provides a pictorial overview of the vast variety of odors in this category (27). The origin alcohol structures are listed horizontally across the top and the origin carboxylic acid structures are in the vertical column. A few important examples are:
>   **Benzyl Acetate** is contained in the essential oils of gardenia and ylang-ylang and produced synthetically for its special fruity, jasmine type aroma;
>   **Benzyl Salicylate** is synthesized to produce fresh azalea and daffodil aromas.

Terpenes

Terpenes are hydrocarbons uniquely synthesized in nature by building molecules in units of five carbon atoms. Many are useful in both perfumery and aromatherapy with a few examples below.
>   **Linalool**, a terpene alcohol, is a natural isolate from coriander and provides a spicy floral scent. It is also known for its antimicrobial, anti-inflammatory, sedative and stress-relieving properties.
>   **Citronellal** is a terpene aldehyde synthesized for its fresh citrus, floral Lily-of-the-Valley and Hyacinth notes.

Additional Chemical Categories and a more comprehensive treatment of the characteristics of synthetics important to perfumery are discussed in the chapter on the Chemistry of Odorants.

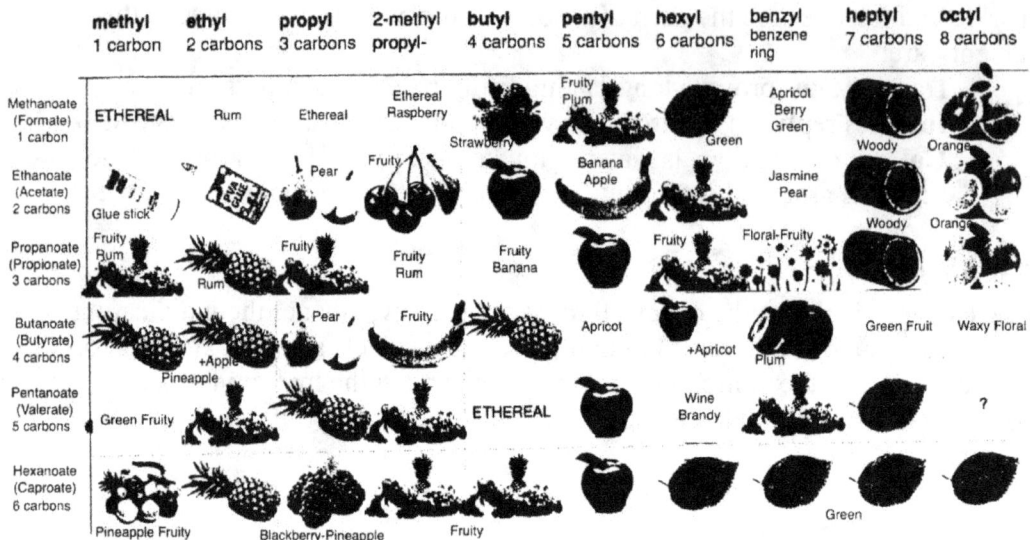

Figure 1.4a Odor of Esters from C1-C5 carboxylic acids (y-axis) and C1-C8 alcohols (upper x axis), modified from Kennedy (jameskennedymonash.wordpress.com).

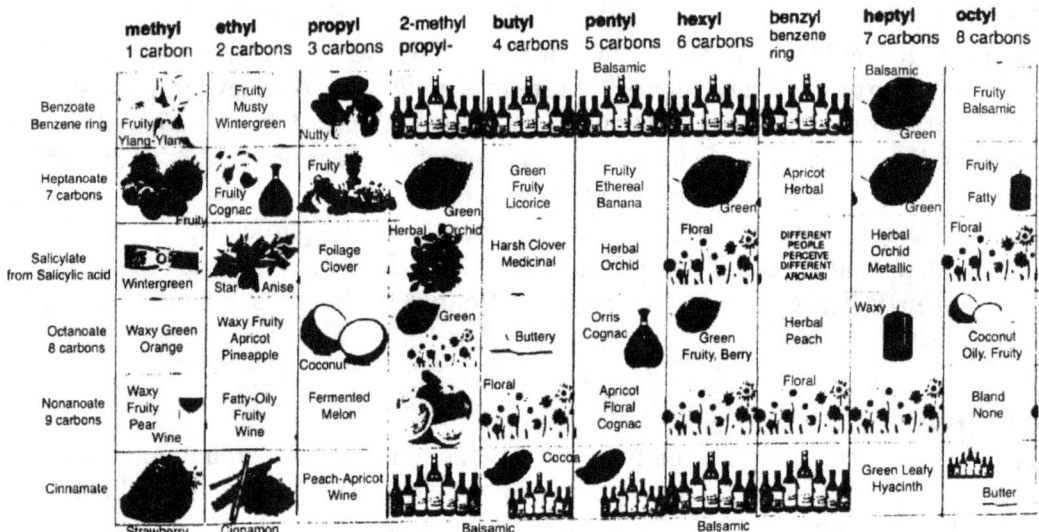

Fig. 1.4b Odors of esters from higher molecular weight acids (y-axis) and C1-C8 alcohols (upper x-axis), modified from Kennedy (jameskennedymonash.wordpress.com).

Chapter 2. Perfumes – History & Modern Perfumery

The use of essential oils and perfumes has a very long history. Plants were early on used for both fragrances and medicinal applications. Practically every aspect of the use of fragrances has been significantly affected by scientific developments right up to modern times. This includes the discovery of distillation techniques, the synthesis on new essences, training methodology for fragrance formulations, gas chromatographic analysis of perfume composition, applications of computer technology and the more recent use of artificial intelligence for creation of perfumes. In this section take a brief look at early history, the emergence of modern perfumery and perfume structure and composition.

Brief History of Perfumes

The origin of use of essential oils in perfumes has been traced back 5,000 years to the Middle East with the Egyptians as skilled perfumers. The formula of an important perfume, *Kyphi* associated with the God Ra, was written in hieroglyphs in the tomb of Tutankhamun (Figure 2.1). There were also vessels in the tomb containing a sweet, resinous, balsamic scent that was still detectable after thousands of years! It is the oldest documented perfume formula and has been recreated in various forms in more recent years (28-35).

Figure 2.1 Tutankhamen's Queen anointing him with perfume in floral pavilion.

Perfume pots and stills from 3,000-4,000 years ago have been found in Cypress and in the Indus Valley of India. Hebrews learned the art of perfume making from the Egyptians and formulations for perfumes can been be found in the ancient Jewish Torah and also from the Orient. In 300 BCE Theophrastus, the Greek botanist, recorded trade of essential oils and perfumes between Phoenicia (Levant, Lebanon), Greece and Arabia.

The first documented evidence of distillation of essential oils was recorded by Herodotus in 425 BCE as a method for distilling turpentine. In 300 BCE Theophrastus' treatise "On Odors" described the basis for blending perfumes, shelf life, use of wine containing aromatics and the effect of odors on the mind and body. Cleopatra is said to have soaked the sails of her royal barge with perfume when she sailed out to meet Mark Antony in 44 BCE (28-35).

The Greeks embraced perfumes and the fragrances they primarily used were of rose, saffron, frankincense, myrrh, violet, spikenard, cinnamon and cedarwood. The Roman, Pliny the Elder, described a primitive technique in AD 77 for collecting essential oil from a resin by condensation on a bed of wool. Another approach was to soak the resin in a vegetable oil and then soak fabric in the fragrant oil for collection of the fragrance (28-35).

Perfumes and fragrances were multipurpose in earlier times, utilized also for personal hygiene as deodorants and antiseptics. Homer extolled their use in bathing and massage for their healing properties, similar to aromatherapy today.

There are also religious connotations with essential oils and perfumes. The Gifts of the Magi were of Frankincense and Myrrh and these two essential oils were items of commerce for 3,000 years. The earthy myrrh was mixed with wine for a drink given to Jesus as a stupefying potion before putting him on cross and thus myrrh is a symbol of suffering and affliction (28-35).

Frankincense, with its woody balsamic scent, was a symbol of holiness and righteousness. It was burned as incense as an offering to God to ward of evil and disease. Indeed, the word perfume derives from Latin word "Fumus," smoke or fragrances diffusing into the air.

In the Muslim world it was passed on that Mohammed filled the mosques with perfumes and musk was incorporated into the mortar during construction. Persian Kings and their courts also made indulgent use of perfumes. An important innovation took place around the year 1020 AD when the Persian philosopher and physician, Avicenna, perfected the art of steam distillation for extraction of essential oils. He produced both rose oil and rose water (floral water) and also discovered alcohol (28-35).

Romans and Greeks had a different perfume for each part of body and Louis XIV personally supervised preparation of a particular perfume for each day of the year.

Perfume leather gloves became very popular in France in the 17th century and both glove and perfumer makers were established in Grasse, France. Grasse eventually became the epicenter for expertise of growing plants for fragrances and collection, extraction and distillation of essential oils for perfumes.

The first fragrance to incorporate alcohol was a potion with a rosemary base made for the Queen of Hungary (Elizabeth of Kujavia, 1305-1380) known as "The Queen of Hungary's Water." This formulation it is considered the precursor to Eau de Cologne so popular for centuries. In the mid-19th century German chemists moved from alchemy to scientific methodology and started producing fragrant synthetic chemicals which revolutionized perfumery (28-35).

Emergence of Modern Perfumery

With the development of synthetics and the perfection of essential oil extraction methods the perfumer's palette was greatly expanded, opening up new potential rich and tantalizing aromas and, in addition, providing more tenacity and stability.

Synthetics in Perfume Formulations

A number of factors have strongly influenced the use of synthetics in perfume formulations over the past roughly 150 years. Although many fragrances are still obtained from natural sources the cost is continually rising due to labor and supply factors. Also, many flowers produce such low concentrations of fragrant essential oils that, for cost reasons, synthetics molecules are typically substituted for the natural aroma. These flowers include freesia, peony, lily-of-the-valley (muget), lilac, mimosa, heliotrope, violet, jonquil, narcissus and hyacinth.

The classic perfumes also often contained musky animal odors for the base notes and as fixatives. However most of these sources are no longer available due to animal scarcity and the cruel, inhumane measures for the oil collection. These musky base notes are now practically all of synthetic origin. A third factor, however, has resulted in elimination of some synthetic chemicals in perfume formulations due to increasing restrictions for health and safety considerations described in a later section.

Before 1870, perfumes were natural and basically literal translations of the flower, for example, the two fragrances *Rose* and *Jasmin* both by Molinard in 1860. By 1930 perfumes already contained up to 15% synthetics, by the 90's the synthetics comprised 85% of the composition and now 80% of perfume fragrances are composed of 95% synthetics. There are now about 1700 synthetic compounds available to perfumery with typically about 750 of these accessible to perfumers in the large perfume houses (13-16).

However, there is also an increasingly strong consumer preference for all-natural fragrances which has created a market for artisanal perfumes. But it is important to note that the use of synthetics opened up a new era of creativity and impressionism in perfumery. The natural oils still constitute the foundation of a perfume providing richness and roundness, while the synthetics provide strength and sparkle and also provide new aromas previously unavailable from natural sources.

Perfume Structure & Composition

Fragrances are sold as a solution of essential oils and synthetics usually diluted in ethyl alcohol in various concentrations as shown below. The concentrations can vary to some extent with different suppliers but generally follow the categories given* (15,16,19,36,37).

Perfume	20%	(15-35%)
Eau de Perfume (EdP)	15%	(10-20%)
Eau de Toilet (EdT)	10%	(5-15%)
Eau de Cologne (EdC)	5%	(3-8%)

* all diluted with alcohol

Perfume Structure

While Piesse in the 1800s is credited with defining terms for perfumery according to musical descriptions as in "notes" and "accords," Jean Carles developed the pyramid classification structure of a perfume for instructional use. The "fragrance pyramid" has proven very useful for description of the basic elements of a perfume and many perfumes comply with this triangular structural configuration. A typical olfactory fragrance pyramid shown in Figure 2.2 has the three phases based on volatility (or vapor pressure) of the notes, Top, Middle and Base notes (15,19,36).

Top or Head Notes are the notes immediately perceived with application of the perfume. They provide the first impression and often consumers will purchase a perfume based only on this initial introduction. The notes evaporate quickly, lasting only 5-15 minutes and are often referred to as fresh, sharp, assertive, of light quality, and immediately apparent. The top notes are usually about 15-25% of the composition. Typical examples include citrus notes of bergamot and lemon or aromatic notes of lavender and coriander.

Middle or Heart Notes emerge as the top notes dissipate. The heart is 30-40% of the blend and can be up to 70%, with these notes providing a more mellow, rounded scent emerging sometime after the first impression. The heart notes provide much of the character and intensity of the fragrance and ground and balance the perfume with the sometimes unpleasant initial impression of the base notes. The middle notes last much longer than the top notes, typically 20-60 minutes. Typical examples include floral notes of rose and jasmine, spicy notes of cinnamon and pepper, or fruity notes of peach and berries.

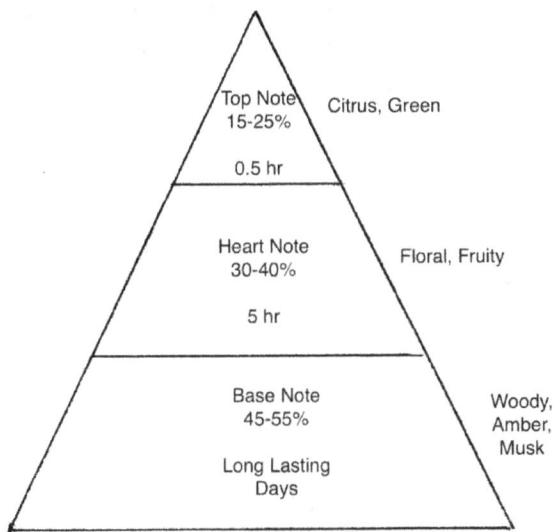

Figure 2.2 Olfactory Pyramid – Triangle of Scents

Base Notes emerge much more slowly, up to 30 minutes after application, and linger for a considerable amount of time, usually more than six hours and even longer. They are sometimes a substantial portion of the fragrance, 45-55%, and provide the foundation to the perfume with their depth and solidity. They are very rich, deep and heavy and impart resonance to the blend. Base notes also perform another important function as a fixative to prevent lighter oils from dispersing too quickly. Typical examples include vanilla, amber, musk, patchouli, oakmoss and woody scents of sandalwood and cedar.

The ratios of the top, middle and base notes vary considerably in different fragrances based on the preferences of the various perfumers. A small sample of the volatility of some top, middle and base notes is given in Table 2.1. There are, of course, many more choices of essential oils and synthetics for use in perfume compostions.

Jean Carles composed a number perfumes based on the pyramidal structure including *Ma Griffe (Carven)*, *L'Air du Temps (Ricci)* and *Cabochard (Gres)*. He is also the perfumer that produced Dana's *Canoe* and *Tabu*, Shirparelli's *Shocking* and *Miss Dior*. Carles' vintage *Ma Griffe* ("My Claw") launched in 1946 is a complex Floral-Chypre perfume that is a true classic fragrance (8,14,15).

The top notes of *Ma Griffe* provide a burst of green and citrus notes with the sparkle of aldehydes (1%). The middle note contains a plethora of florals, but with muted sweetness, from a traditional combination of jasmine and rose, muget from hydroxycitronellal (10%), iris, ylang-ylang and a hint of gardenia. It settles into a heart of a classic chypre accord, a mossy floral with spice and musk, composed of oakmoss, vetiveryl acetate (10%), labdanum, styrallyl acetate (4%), styrax, methyl ionone, aldehyde C14, coumarin, patchouli, sandalwood, musk ketone and amber. The latter amber note from a combination of vanilla, labdanum and olibanum. The perfume stays green, but the dry down is very soft and smooth. An engrossing herbaceous French perfume (15,32,33).

Table 2.1 Examples of Perfume Note Volatility

Top Very Volatile EC*= 1-14	Middle (Modifiers) Intermed. Volatility EC*=15-60	Base Low Volatility EC*=61-100
Lemon (Grass)*	Paraguay Petitgrain	Orange flower, abs
Bergamot*	Galbanum	Syrax
Grapefruit	Juniper	Oakmoss, abs*
Lime*	Cloves	Frankincense
Anise	Neroli*	Myrrh
Coriander	Black pepper	Rosewood
Lavender*	Rose abs*	Vetiver
Eucalyptus	Jasmine abs*	Patchouli*
Methyl salicylate	Ylang-Ylang*	Sandalwood
Amyl acetate*	Terpineol	Vanilla*
Phenylethyl acetate	Phenylethyl alcohol	Peru Balsam
Linalool	Methyl anthranilate	Santalol (Santal*)
Benzyl cinnamate	Geraniol	Ionones
Linalyl acetate *	Methyl heptin carbonate	Musks*
Bornyl acetate	Citronella	Phenylacetaldehyde
Anisaldehyde	Eugenol*	Benzyl salicylate
Aldehyde C8, C9	Aldehyde C11, C12	Aldehyde C16, C18

*EC=Evaporation Coefficient - a relative estimate of the evaporation times of the oils & synthetics based on the time that it takes for the odor to disappear from 100 mg on smelling strips. References: Poucher et al. (25), Carles (38, 39)

Notes, Accords and Bases

It is important at this juncture to differentiate several terms used when describing the composition of a perfume, namely **Notes, Accords, Families, Facets and Bases** (15,16,19,36,40-44). In the 19th century the Frenchman Piesse classified the odors of essential oils according to musical scales, thus **Notes** are descriptors of the scents in a perfume, similar to tones or notes in music, and are the fundamental components utilized for descriptions in perfumery. They are often distinguished as top (most volatile), middle (intermediary volatility) and base (lowest volatility) notes in the structure of a perfume as previously described.

An **Accord** (or **Family**) is a new or familiar scent made up of usually three or four notes blended together to form a distinct fragrance, again similar to music terminology where notes are combined to form a single cord. A **Facet** is an olfactory feature of a perfume and the term deviates from musical comparisons; it is more related to classification of gems. The main olfactory **Facet** determines to which **Family** the perfume is categorized, while less intense Facets in the perfume define the originality of the fragrance. Thus, Perfumes are multifaceted and Facets are similar to Accords with considerable overlap; for example, the Floral category can be an Accord, Family and a Facet (15,16,19).

Perfumers also refer to **Bases**, not necessarily base notes as defined above. **Bases** are preformed "blocks" of scents that can be as simple as an **Accord** or as complex as a complete perfume; they are a type of modular perfume. Bases can be used in a similar way as essential oils for creating a fragrance.

As an essential structure of a perfume, **Bases** have a specific character. Perfumers can create their own unique **Bases** for construction of signature perfumes or obtain any number from perfume supply houses. The bases can be formulated from the composition of the headspace of floral notes that are not obtainable due to the low contents of essential oils in the plant, such as hyacinth and gardenia, or specialty odors bases can be created such as "fresh cut grass" or "sweet fruity plum" or even "cola" (15,16,19).

Bases are mixtures of chemicals as are essential oils. Many perfumers use **Bases** for construction of perfumes while others use a more "free-lance" approach using the **Bases** for special effects, the latter more reflective of modern perfumes of greater simplicity.

Perfume Accords

A Perfume Accord is comprised of different notes that blend together so well, you smell a harmonious odor. An example of this concept is a recipe for a pleasant fruit aroma by combining discordant ingredients of Butter, Fresh Grass, Ripe Apple & Cotton Candy that together provide the aroma of Strawberry! Let us now review the fragrance accords in more detail.

The typical **Traditional Accords** in perfumery are Chypre, Fougère, Oriental or Amber, Floral, Woody and Leather. More **Modern Accords** include Citrus, Green, Aquatic (Oceanic/Ozonic), Fruity and Gourmand. Some of these later families are grouped together under a common thematic accord such as **Fresh** for Citrus, Green, Aquatic and Aromatic. There are also many subgroups for each of these families and considerable overlap of accords and subgroups. The Fragrance Wheel (Figure 2.3) further described in a following section provides a simplified overview of the relationship of the families or accords of perfumes. In the discussion below an example perfume is described in more detail for a few major accords (15,16,40-44).

Traditional Accords

Chypre Accord

The Chypre accord is based on the archetype perfume *Chypre* composed by Francois Coty in 1917. It is a classical structure with the freshness of the citrus bergamot for the top note with jasmine flower, the resinous extract labdanum, the intense fragrance of oakmoss, and some animal civet and musk. The aroma progression is from smoothly sour to sweet and resinous, drying down to earthy green-moss, a harmonious accord of multiple parts. Patchouli, vetiveryl acetate and methyl ionone are also incorporated into the base in later versions.

A wide range of fragrances in this accord have been produced by addition of notes of aldehydes, flowers, fruits, grasses, leather, etc. such that this is one of the major categories of perfumes along with Florals and Orientals. Subgroupings are Chypre Floral, Fruity, Green and Leather. Some other significant fragrances in this accord include, *Ma Griffe (Carven), Miss Dior and Cabochard (Gres), Femme (Rochas), Channel 19 and Mitsouko (Guerlain)* (15-17,45,46).

Mitsouko is a mysterious fragrance, cool top notes with an oak moss base. But it also has a juicy, peach gourmand nuance. The well balanced fragrance is exuberant, unusual and elegant. It was created in 1919 by Jacques Guerlain and inspired by the novel *La Bataille* (1909, "The Battle") by Claude Farrère. It is a story of the impossible love between Mitsouko, wife of Japanese Admiral Togo, and a British officer, but there is really nothing Japanese about the perfume (32,33).

As a prime representative of the Chypre accord, the base note of *Mitsouko* is composed of vetiver oil (10%), benzoin, coumarin, methyl ionones, clove oil and oakmoss. The percentage of vetiver oil is quite high for a feminine fragrance and it also contains 0.4% labdanum resinoid. The top notes include, in addition to bergamot, γ-undecalactone which provides the peach aroma. Due to IFRA health regulations, more recent formulations of Mitsouko and other Chypre and Fougère accord fragrances have Orcinyl 3 or Evernyl substituted for Oakmoss.

A pyramid structure given for this fragrance is (11,12):
Top –   Bergamot, Citruses, Peach
Heart -   Jasmine, Rose, Ylang-Ylang, Lilac
Bottom -   Oakmoss, Vetiver, Cinnamon (spices)
Fragrance – Woody, warm spicy, earthy, citrus, cinnamon, floral.

Fougère Accord

Fougère is Fern in French and the classical structure is based on lavender, oakmoss and coumarin with bergamot, geranium and vetiver. The name originated with the *Fougère Royale* perfume created by Paul Parquet for Houbigant. It is closely related to the Chypre accord but with a "fresher" olfactory appeal, especially for masculine fragrances. The more modern formulations of Fougère also include citrus, herbaceous green notes, and anamalic and floral notes. Additional perfumes in this family include *Brut (Faberge), Drakkar Noir (Guy Laroche), Douro (Penhaligon)* (15-17,45,46).

Oriental/Amber Accord

This accord employs a reference to the Middle East as mysterious and sultry, typified by rich softness. The classic structure is based on vanilla with gum resins, patchouli, sandalwood, heliotrope, coumarin and orris. The accord also often contains floral, spicy and amber notes. The latter amber component is used to provide a foundation for the accord; however, there is no such thing as an amber essential oil and it is not related to the amber from fossilized tree resin typically utilized for amber jewelry. An amber base is created by a fusion of balsamic notes such as vanilla, cistus labdanum, benzoin and Tolu balsam as sweet, warm resins.

Oriental fragrances are exotic and evocative for both men and women. The primary subgroups for this accord are Floral Woody, Floral Spicy, and Soft- and Semi-Oriental. Fragrances in this category include *Shalimar (Guerlain), Opium (St. Laurent), Mademoiselle, Coco (Chanel), Youth Dew (Lauder), Obsession (Calvin Klein) and Old Spice (Shulton)* for men (15-17,45,46).

Shalimer is one of the best-selling oriental perfumes of all times. It is sensual and voluptuous, velvety, plush, undeniably uniquely sweet and penetrating. It is named after the favorite gardens of Mumtaz, wife of Emperor Shahjahan, who built the Taj Mahal in her memory. Jacques Guerlain introduced the fragrance in 1925 and it was reissued in 2007 (11,12,32,33).

A pyramid structure for this fragrance is (11,12):
Top – Lemon, Orange, Cedar
Heart - Jasmine, Rose, Rosewood
Bottom - Sandalwood, Vanilla, Vetiver, Incense
A civet note is a dominant facet in *Shalimer* as well as in the classic *Jicky* fragrance.
Fragrance - Balsamic, citrus, powdery, woody, vanilla, smoky

Floral Accord

This floral bouquet accord accounts for many sweet, soft and gentle fragrances. It is the largest family and often easy to recognize; typically distinguished by the heady scent of blooming flowers of jasmine, rose and ylang-ylang, sometimes accompanied by a violet scent from ionone. These fragrances have a strong feminine appeal.

Floral bouquets include the "White Florals" that are known for their mesmerizing power. A bouquet of white florals combines narcotic scents of jasmine, orange blossom, lilac, gardenia, jasmine and lily-of-the-valley. However, many florals have a dominating floral scent such as the distinctive carnation in *L-Heure Bleue (Guerlain)*. There are several additional subgroups such as Aldehydic Floral and Green/Fresh Floral.

There are a host of perfumes in this category with **the most famous *Chanel No. 5*** and others that focus on a particular flower or bouquet of flowers such as *Fracas (Robert Piquet)*, **Quelques Fleurs (Houbigant)**, *Joy (Jean Patou)* , *Paris* (rose, *Yves St Lauren) Dorissimo* (lily-of-the-valley, white florals, *Dior)*, *Muget (Molinard), Beautiful (Lauder), Chamade* (hyacinth, *Guerlain), Narcisse Blanc (Caron), Apres l'Ondée* (violet, *Guerlain), Hiris* (iris, *Hermès), L'Air du Temps* (carnation, *Nina Ricci), Savage Jasmine (Sana Jardin), Pour un Homme* (lavender, *Caron), Anaïs Anaïs* (lily, *Cacharel), Fracas* (tuberose, *Robert Piguet), Tiare* (tiare flower, *Chantecaille), Frangipane (Chantecaille), Gardenia (Roja Dove), Coeur d'Ylang (Comptoir Sud Pacifique)* and *Lilac (Roja Dove)* (15-17,45,46).

Chanel No. 5 was produced in response to the famous Coco Chanel requesting that "A woman should smell like a woman and not a Flower." She wanted a sexy, provocative fragrance but with a clean aspect. Coco had a rather inauspicious start since she was orphaned at 12 years old, raised in a convent where she learned to sew and later developed her famous brands of clothing as well as fine fragrances (15-17,28-35).

Chanel No. 5 was launched in 1921 and is the best-selling perfume in history, resulting in the new subgroup, Floral-Aldehyde perfumes. The perfumer, Ernest Beaux, is credited with applying a high dose of aldehydes to the composition, but another version of the discovery indicates that his assistant overdosed the juice by mistake. But it is, indeed, the aldehydes that provide the special sparkle and freshness to this spectacular fragrance.

The composition is a monument to perfect structure and texture with a total of 80 ingredients. The iconic perfume provides the concept of "Clean Sheets & Warm Bodies" as a powdery, luxurious signature fragrance. When Marilyn Monroe asked what she wore to bed at night, she replied, "Nothing but *Chanel No. 5*" (28-35).

Notes identified in the pyramid structure of *Chanel No. 5* include (11,12):
Top -   Aldehydes, Bergamot, Ylang-Ylang, Neroli
Heart -   Jasmine, Rose
Bottom -   Sandalwood, Vanilla, Vetiver
Fragrance - Floral, powdery, woody, citrus, vanilla

Some further constituents in the composition of *Chanel No. 5*, many in the original formulation, include (45-57):
0.6%   Aldehydes, C10/C11/12, 1:1:1
5%   Bergamot oil
1%   Petitgrain oil
4%   Jasmine absolute
0.5%   Rose oil
0.5%   Java vetiver oil
3.5%   Musk ketone
2.5%   Musk ambrette
15%   Natural Tonquin musk tincture
15%   Natural civet infusion

Fracas is an intoxicating Parisian style floral fragrance launched in 1948 by Robert Piquet and reintroduced in 1998. The perfumer Germaine Cellier produced this glamorous and superbly feminine fragrance. The dramatic notes of tuberose in the composition provide for a very sexy and intoxicating fragrance. The perfume opens with sweet, buttery, citrus notes, then a fusion of white flowers with a dominant note of tuberose, embraced with musk and oakmoss, with a touch of iris in drawdown.

Notes identified in the pyramid structure of *Fracas* include (11,12,28-35):
Top -   Orange Blossom, Bergamot
Heart -   Tuberose, Jasmine, Green Leaves
Bottom -   Musk, Amber, Oakmoss, Vetiver
Fragrance - White floral, tuberose, floral, animalic, green, woody

Woody Accord

Fragrances that are dominated by woody scents, typically of sandalwood and cedar. Patchouli, with its aromatic camphoraceous smell, and vetiver are also commonly found in these perfumes. Perfumes in this family include *Sahara Noir (Tom Ford), Cashmere (Luigi Borrelli), Pearl (La Rive), Santal (Fragonard), Bel Canto (Galimard), Bois Fonce (Ava Luxe), Dior Homme (Dior), Vetiver (Guerlain), Bois des Îles (Chanel), Rumba (Balenciaga), Égoïste (Chanel)* and others more thematic as *Oud Extrait (Maison Francis Kurkdjian)* and *Sandalwood (Demeter)* (15-17,45,46).

Burberry Body – A woody fragrance that arrives on the market in 2011 in over 150 countries worldwide.
Notes identified in the pyramid structure of the fragrance include (11,12,28-35):
Top -   Peach, Freesia, Wormwood
Heart -   Rose, Iris, Sandalwood
Bottom -   Musk, Amber, Vanilla, Cashmere Wood
Fragrance – Woody, musky, powdery, rose, floral

Leather Accord

These fragrances are warm and opulent, very masculine and evoke smoky notes of tobacco, wood and wood tars. The scent insinuates leather that often depends on synthetic quinolones synthesized in the 1880's. Subgroups in this family are Mossy Woods and Dry Woods. Some perfumes in this category include *Bandit (Robert Piguet), Cuir de Russie (Chanel), Cabochard (Gres), Joop! (Joop!), Electric Heat (Joop!), Jolie Madame (Pierre Balmain), A*Men Pure Leather (Thierry Mugler)* (15-17,45,46).

A*Men Pure Leather was launched by Thierry Mugler in 2012. This concept combines haute perfumery and the traditional craft of glove making and perfume craft. The uniqueness of this fragrance is that it was made by soaking the original compositions in natural leather for four weeks. It exudes dark and fiery overtones providing the fragrance with an irresistible magnetism of mint, coffee, carmel and chocolate (11,12,32,33).

Notes identified in the pyramid structure of the fragrance include (11,12,28-35):
Top – Coriander, Lavender, Peppermint, Bergamot
Heart - Coffee, Atlas Cedar, Leather, Patchouli
Base - White Musk, Tonka Bean, Carmel, Chocolate, Vanilla

Modern Accords

Many new accords were developed as new synthetic chemicals with unique odors were created in perfume houses or in independent firms. The Citrus, Green, Aromatic and Aquatic Accords introduced below are sometimes collated under an encompassing "**Fresh Accord**" but are described separately in this synopsis.

Citrus Accord

An older fragrance family formerly comprised of fresh eau de colognes due to the volatility of citrus scents. Development of new synthetic compounds has allowed for the creation of more tenacious citrus fragrances.
These perfumes are characterized by their zesty freshness and lightness. Perfumes

in this family include *Lime Basil & Mandarin (Jo Malone), Velvet Bergamot (Dolce&Gabbana), Mediterraneo* (lemon, *Carthusia*), *Eau Savage* (lemon, *Dior*), *Pamplelune* (grapefruit, *Guerlain*), *Aqua Allegoria Limon* (lemon, *Guerlain*) and *Tangerine Vert (Miller Harris)* (15-17,45,46).

Aromatic Notes can be added to enhance the citrus accord including, for example, thyme, rosemary, tarragon or mint. Examples include: *CK One (Calvin Klein), Eau de Rochas (Rochas)*, and *O de Lancome (Lancome)* (15-17).

Lime Basil & Mandarin is a bracing rich citrus fragrance with an intriguing grassy mint smell of basil. Jo Malone launched this clear and bright, aromatic citrus perfume in 1999. The feel is of a natural, clean, relaxed mid-morning shower.

Notes identified in the pyramid structure of the fragrance include (11,12,28-35):
Top –   Lime, Mandarin Orange, Bergamot
Heart –   Lilac, Iris, Basil
Bottom -   Patchouli, Vetiver
Fragrance – Citrus, fresh spicy, green, aromatic, herbal, woody

Green Accord

A lighter, more modern interpretation of the Chypre type, with pronounced cut grass, crushed green leaf and cucumber-like scents. Galbanum is a typical ingredient in this type of perfume. Incorporation of the synthetic cis-hexenol imparts the cut green grass odor and nonadienol provides a violet/cucumber accent. Some perfumes from this family include *Alliage (Lauder), Eternity (Calvin Klein), Vent Vert (Balmain), Homme Nature (Yves Rocher), Água Verde (Salvador Dali);* more spicy is *Lauren (Ralph Lauren)* (15-17,45,46).

Aquatic (Oceanic/Ozonic) Accord

Another newer category appearing in 1991. A very clean, modern smell leading to many of the contemporary androgynous perfumes. This family of fragrances suggest the aromas of the ocean, sea breezes, waterfalls and mountain air. They typically contain citrus odors and the synthetic, calone, discovered in 1966 which has a sea breeze scent with floral overtones. Perfumes in this category include *Cool Water (Davidoff), Acqua di Gioia (Armani), Pour Homme (Kenzo), Nebbia Spessa (Nebbia), Aria di Mare (Il Profumo), Eternity, Dune (Dior), Acqua (Jeanne en Provence)* and *L'eau d'Issey (Issey Miyake)* (15-17,45,46).

Cool Water emanates fresh, aromatic aquatic notes that recall ocean breezes and cool sea-water. The perfumer Pierre Bourdan combined minty green notes with lavender notes of sea water in the top, classic flowers in the heart rounded out in the base with woods and oakmoss for a very masculine fragrance creation.

Notes identified in the pyramid structure of the fragrance include (11,12,28-35):
Top –   Lemon, Melon, Lotus, Calone
Heart -   Jasmine, Water Lily, Lily-of-Valley
Bottom -   Sandalwood, Musk, Vetiver
Fragrance – Aquatic, floral, fruity, ozonic, sweet, fresh.

Cool water also contains 0.7% Egyptian geranium oil, 3.5% lavandin oil and 9.5% synthetic linalyl acetate. This fragrance is also classified as a fougère perfume.

Fruity Accord

This is a newer family featuring fruits other than citrus, such as peach, black currant, mango, pineapple, passion fruit and others. This accord relies heavily on synthetic materials to create the desired aroma (see Chapter 4). Examples of this fragrance are *Baby Doll* (black currant, pineapple, *Yves St. Laurent*), *Pulp* (black currant, fig, *Byredo*), *La Danza delle Libellule* (apple, strawberry, *Nobile*), *Péché Cardinal* (peach, *MDCI*), *Lost Cherry (Tom Ford), Premier Figuier* (fig, *L'Artisan*), *Lady Boy* (banana, *Gorilla*), *Mango Manga (Montale)*, 1804 (pineapple, *Histoires de Parfums*), *Botrytis* (honey, pineapple, *Ginester*) (15-17,45,46).

Baby Doll is a fragrance that is both youthful and playful as well as elegant and fresh. A Fragrance Foundation Fifi Award winner, it is a fruity, floral fragrance with lots of gourmand notes. The perfumers Cecile Matton and Ralf Schwieger created this "Happiness in a Bottle" for Yves St. Laurent (11).

Notes identified in the pyramid structure of the fragrance include (11,12).
Top -   Black Currant, Orange, Apple, Pineapple
Heart - Rose, Freesia, lily-of-the-Valley, Heliotrope
Base - Cedar, Sandalwood, Tonka Bean, Vanilla

Gourmand Accord

This family includes perfumes that have "edible" type notes of chocolate, vanilla, carmel, honey and candy with base notes of patchouli and musk. The gourmand trend has steadily increased in popularity since 1992 with the launch of Thierry Muglers' perfume, *Angel,* featuring a "cotton candy" type aroma from the synthetic ethyl maltol. However, it is not a totally new approach as Edmond Roundnitska created D*iorissimo* in 1956 with the heavy and sweet notes found in today's gourmand perfumes. This family also includes perfumes *A*Men (Mugler), Euphoria (Calvin Klein), Rochas Man (Rochas), Antidote (Viktor&Rolf), Crême Brûlée (Zara), Crême de Pistache (Laura Mercier), Marrons Glacés (Laura Mercier)* and *Juicy Couture (Juicy Couture)* (15-17,45,46).

Juicy Couture checks off to a lot of categories of families as it combines Floral,

Fruity and Gourmand scents into an instant hit with both American and European consumers, and perfume critics as well. Launched by Juicy Couture in 2006, it became another Fragrance Foundation FiFi Award winner.

Notes identified in the pyramid structure of the fragrance include (11,12).
Top -   Watermelon, Passion Fruit, Hiacynth, Green Apple & Leaves, Madarin, Orange and Marigold
Heart -   Tuberose, Lily and Rose Hips
Bottom -   Caramel, Crème Brulee, Vanilla, Patchouli and Precious Woods
Fragrance - sweet, white floral, green, caramel, vanilla, fruity, aquatic

Fragrance Perception

Zarzo and Stanton (48) compiled a list of odor descriptions from the literature for a range of natural oils and synthetics and some examples are shown in Table 2.2. They gave descriptions for thirty reference materials based on the perceptions of a panel of people. Their list provides a useful reference for odor descriptions of the different attributes assigned to the various essential oils and chemicals and serves as a starting point for development of fragrance classification schemes.

Table 2.2 Odor Descriptions for Some Natural and Synthetic Materials

Category	Reference Material	Odor Description
Aldehyde	Aldehyde C10	Sweet, citrus peel, touch rancid-fatty
Aromatic	Vanillin	Sweet, vanilla, chocolate, balsamic
Balsamic	Oilbanum	Balsamic, spicy, coniferous, resin, lemon
Fatty	Undecylenic alcohol	Soapy, waxy, floral, rosy pleasant
Floral	Jasmine, abs	Honey-like sweet floral, fruity-herbaceous
Fresh	Bergamot oil	Fresh, clear, lively, fruity sweet citrus
Fruity	Hexadecanal	Strong strawberry odor
Green	Methyl octyonate	Fresh leafy, floral violet odor
Lavender	Lavender oil	Dry, fresh balsamic, herbaceous, floral
Powdery	Mixture-Musk ketone	Warm, sweet, musky
	+ Coumarin	Sweet, herbaceous, spicy, fresh cut hay
Spicy	Eugenol	Warm spicy, medicinal, cloves, balsamic
Sweet	Heliotropin	Warm, sweet, floral narcotic, cherry pie
Watery	Cyclamen aldehyde	Floral green, watermelon, fresh rhubarb

Reference: Zarzo & Stanton (48) and references cited therein

With about 500 new fragrance launches per year it is valuable to have an accurate classification system based on perfume smell (23,35,49-51). A number of classification schemes have been developed over the years by perfume houses for this purpose (40,52,53). The diagrams have been in the form of hexagons or more commonly circular schemes such as the Fragrance Wheel shown in Figure 2.3

which classifies fragrances in 14 categories around a central hub (53).

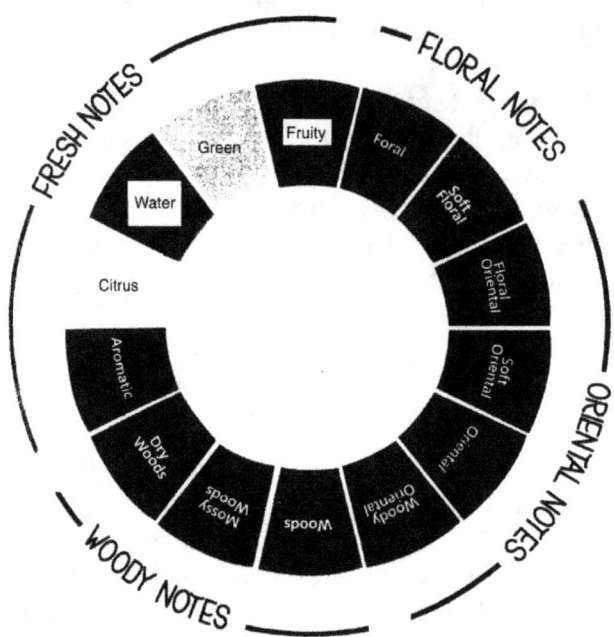

Figure 2.3 Fragrance Wheel (53).

The Fragrance Wheel has proven valuable in the fragrance industry, especially for retail sales. There have been several versions proposed in the past but this recent one, designed by perfume consultant Michael Edwards, was updated in 2010. The circular display relates the different scent families and subfamilies in a systematic pattern. The scents are grouped to show the relationship based on similarities and differences. Scent groups that border each other share common olfactory characteristics while those further away are not as strongly related. Selection of related perfumes can be more easily made by identifying the Family or Accord as Floral, Oriental, Fresh or Woody and then further identifying the related subgroups.

Further work on the relationship among odor descriptors used in perfumery was carried out by Zarzo and Stanton (48). In this excellent study their goal was to develop a sensory map with the common perfume odor descriptors. They analyzed two odor databases of perfume materials using a Multivariate Statistical Method called Principal Component Analysis. The databases were the Boelens-Haring numeric database (rating of test odor compared to standard) and Thiboud's semantic odor profiles (words to describe a smell). They then overlaid the results of their two-dimensional sensory map of odor descriptors from the rigorous statistical analyzes onto the Odor Effects Diagram previously proposed by Jellinek in 1951 (52). A version of this overlay is shown in Figure 2.4 as composed by Donna (54).

The four key effects of erogenous vs anti-erogenous and narcotic vs stimulating at opposite ends of the two axes from Jellinek's diagram are shown in Figure 2.4.

Allowing for some adjustments, there is striking conformity for the positions of the descriptors in the combined composition. Based on the various perfumery databases and odor maps, Zarzo and Stanton provided useful associations among odor descriptors for neighboring smells on the odor map as follows (48,54):

Smoky – burnt, birch tar, toasted, leather
Camphoraceous – pine, lavender, mint, conifer, rosemary
Herbaceous – chamomile, lavender, rosemary, sage, clary sage
Resinous – olibanum, tree gum, conifer
Earthy – dust, moss, forest, soil, mold, must, roots, yeast, mushrooms
Sweet – balsam, vanilla, heliotropin, honey, syrup

The results of Zarzo and Stanton's analysis also essentially verifies the Fragrance Wheel developed by Edwards from his extensive "Fragrances of the World" database (48-54).

Figure 2.4 Overlay of 2-D sensory map of Odor Descriptions of Zarzo & Stanton (48) with Odor Effects Diagram of Jellinek (52). A: Erogenous, alkaline, animal (rich, musk, vanilla); B: Anti-Erogenous, acid, sour, refreshing (citrusy). Circular dots are the odor descriptors and star dots are from the Boelens-Haring database. The mid-section of lower right quadrant was labeled "herbaceous" and the mid-section of the lower left quadrant was labeled "woody" in the corresponding simplified diagram of Calkin & Jellinek (15); (Donna, 54).

Donna (54) also extracted data on feminine versus masculine fragrances from Zarzo and Stanton's study using Edward's fourteen subfamilies. She reported that 42% of women's fragrances are Floral versus just 1% of men's fragrances and middle notes are typically 96% floral for women vs 67% for men. Another comparison was made for the Woody family which accounts for 15% of men's fragrances versus just 2% for women perfumes. Fresh scents showed twice as much in men's fragrances at 14% compared to women's at 7%. Also notable is the 96% of Fresh top notes for men's fragrances versus 52% for women's perfumes (54).

Timeline of Modern Perfumery

We will now follow the progression of modern perfumery, highlighting some of the classical and more significant perfumes, through to contemporary times and the use of synthetics as noted for their initial introduction in Table 2.3 (13-15,19,35,55). Perfumes are noted in italics as well as synthetics in their formulation. The incorporation of synthetics in fragrance formulations has increased dramatically in more recent creations and a list of some of many of the important synthetic compounds used in perfumery and their odors is provided in Appendix I (28-33,45,46,55,56).

Table 2.3 Timeline of Some Synthetics in Modern Perfumery

Year	Perfume	House	Synthetic Compound
1882	Fougère Royal	Houbigant	Coumarin
1889	Jicky	Guerlain	Vanillin & Linalool
1893	Vera Violetta	Roger&Gallet	α- & β-Ionone
1898	Trèfle Incarnat	L.T. Piver	Isoamyl salicylate
1904	La Rose Jacqueminot	Coty	Rhodinol (L-citronellol)
1905	L'Origan	Coty	Heliotropin, Vetiveryl acetate
1906	Après l'Ondée	Guerlain	p-Anisaldehyde
1913	Quelques Fleurs	Houbigant	Hydroxycitronellal & Aldehyde C12 (MNA)
1919	Mitsouko	Guerlain	γ-Undecalactone
1921	Chanel No.5	Chanel	Aldehydes C10-C11, Musk ketone
1922	Nuit de Noel	Caron	6-Isobutylquinoline
1932	Tabu	Dana	Phenylethyl acetate, Civettone
1946	Ma Griffe	Carven	Styralyl acetate
1953	Youth Dew	Lauder	Amyl acetate, Cedryl acetate
1966	Fidji	Guy Laroche	PTBCHA, cis-3-Hexyl salicylate
	Eau Savage	Dior	Hedione (methyl-dihydrojasmonate)
1978	Mure et Musc	L'Artisan Parfumeur	Helvetolide
1979	Nahéma	Guerlain	α-Damascone
1982	Drakkar Noir	Guy Laroche	Ambrox
1985	Obsession	Calvin Klein	Ethylene brassylate
1988	Cool Water	Davidoff	Dihydromyrcenol, Allyl amyl glycolate, Synthetic musks
	Fahrenheit	Dior	Methyl heptine carboxylate (MHC)
	New West for Him	Aramis	Calone
1990	Trésor	Lancome	Galaxolide & Iso E Super
1992	Angel	Mugler	Ethyl Maltol
1999	J'Adore	Dior	Macrocyclic musks

References: Ohloff et al. (13), Vosnaki (19), Nicolaï (35), Pybus (55).

Late 1800s

The beginning of modern perfumery can be traced to the creation in 1882 of *Fougère Royale* by Paul Parquet for Houbigant. He incorporated synthetic *coumarin* with the hay-like aroma based on the natural isolate from Tonka Bean. During this important decade for perfumery the hugely popular perfume *Jicky* was created by Aimé Guerlain in 1889 by the incorporation of *vanillin* from vanilla into the formulation. At the time *vanillin* was produced as a semi-synthetic compound from eugenol oil. *Jicky* is the first truly "vertical" fragrance with distinct top, middle and base notes. Despite the early history of this fragrance, it is timeless and modern. The aroma is lavender-vanilla with citrus-herbal notes and a formerly civet base note (28-33,35-55). A vintage advertisement from for synthetic scents is shown in Figure 2.5.

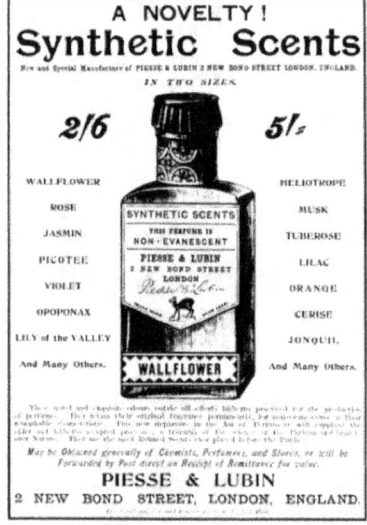

Figure 2.5 Vintage perfume advertisement from Piesse & Lubin in 1890 in London.

Early 1900s

In 1920 the law was passed giving women the right to vote and it was during this period that they redefined themselves. It was the "Jazz Age" of hedonism and the Flappers, "Les Années Folles," the Crazy Years (11,19,28-33,35,45,46,55).

Some of the truly great classic perfumes were composed during the period of 1900-1920. It was a new century and more synthetic materials were becoming available to the perfumer including *coumarin, heliotropin, ambreine, undecalactone* and *aldehydes (C10-C12)*. It was during the period that the *Chypre* accord originated and some of the most famous perfumers realized their creative spirit including François Coty, Jacques Guerlain and Ernest Beaux. In 1922 Coco Chanel proclaimed "A woman should smell like a woman not like a flower" furthering the revolution in modern perfumery.

In 1905 François Coty, a very successful visionary perfumer, created *L'Origan*, a floral fragrance in which he incorporated several synthetics of *coumarin*, as well as the violet aroma of *ionones, heliotropin (piperonal)* which simulates the powdery, almond-vanilla nuances of heliotrope, and *vanillin* of vanilla to produce the spicy, powdery woody fragrant perfume. Coty was very successful with his approach to perfumery as he described: "Give a woman the best product you can compound, present it in a container of simple, but impeccable taste, charge a reasonable price for it, and great business will arise such as the world has never seen." Coty continued his prodigious work by formulating *Chypre* in 1917 with leathery synthetic *quinolines*.

Jacques Guerlain also utilized *heliotropin* and a note of natural vanilla to create the sublime floral mimosa-type perfume *Après l'Ondèe* in 1906. He then created *L'Heure Blue* in 1912 which is a mysterious, elegant perfume with florals of rose and carnation and synthetics of *ionones* and *heliotropin*. *Quelques Fluers*, introduced by Houbigant in 1913, has been an inspiration to many perfumers for the intense floral bouquet. The perfume was balanced by incorporation of a recently available synthetic at the time, *hydroxycitronellal*, which also provided a fresh, green, delicate muget aroma to the composition.

During this period Guerlain created two classic, timeless perfumes both of which are some of the best all time selling perfumes, *Mitsouko* in 1919, a fragrance that is unusual, exuberant and elegant and *Shalimar* in 1925, a magical luxurious perfume. The peachy aroma in *Mitsouko* is attributed to the incorporation of the synthetic, γ-*undecalactone*.

But it was in 1922/1925, that the block buster perfume, *Chanel No.5*, was launched by Coco Chanel and the Wertheimer Group. It has been reported that the perfumer, Ernest Beaux, added "considerably extra" synthetic *aldehydes (C10-C12)* to the formulations leading to the exuberant metallic, shimmering sparkle of the perfume. It became, and still is, one of the best-selling perfumes of all time and, with Coco Chanel involved, fashion and fragrance became inextricably linked. She stated that "No elegance is possible without perfume. It is the unseen, unforgettable, ultimate accessory."

Inspired by the success of *Chanel No.5*, Jeanne Lavin followed with another floral-aldehyde perfume, the spectacular *Arpège* in 1927 composed by perfumers Andrè Fraysse and Paul Vacher. The fragrance was aptly named by Andrè's daughter based on a the musical arpeggio chord. Along with the aldehydes in the top note there is peach, muget & bergamot, a host of florals as middle notes and a classic base including sandalwood, amber, musk and vanilla.

I spotted the striking cobalt blue flagon of *Evening in Paris* at a flea market some years ago and immediately grabbed it up. It was originally launched in 1928 as *Soir de Paris* by Bourjois and is another composition by Ernest Beaux. The name reflected the gaiety and romance of Paris and became very popular in the USA starting in 1929. The immediate aroma from the fragrance is the strong vintage violet smell with peach, apricot and bergamot. The heart note nicely compliments with a dominant iris aroma along with heliotrope, classic rose-jasmine and ylang-ylang. The base notes are amber, sandalwood, vanilla and musk.

A dramatic increase in perfume launches were realized after the outstanding success of the classic perfumes created in the 1900s. The number of new creations have been steadily increasing over the decades, especially after WWII. In recent decades there have been hundreds of new perfumes launched onto the world market. Some significant perfumes created for each of the subsequent decades are briefly described in the following sections. Most of the perfumes selected for this synopsis were outstanding creations or very popular in terms of sales.

Of course, with the profusion of perfumes created since the late 1800's it is impossible to cover even all of the best perfumes, but an attempt is made to cover some personal choices.

The 1930s

The stock market crash of 1929 and the Great Depression led to this more sedate decade. The haves and the have nots became starkly more apparent. Women's hems went down and their hair went up. As sense of idealism pervaded with casual sports clothes apparent during the period (11,19,28-33,35,45,46,55).

Creativity in perfumery continued its march into the 1930's with creation of several more classic perfumes. Additional synthetics such as *phenylethyl acetate* and *civetone* began readily available during this decade. The perfume *Joy* was launched in 1930 by Jean Patou composed by perfumer Henri Almeras. The perfume had a heavy concentration of absolutes of rose and jasmine with the result that it was the most expensive perfume at the time. With the strong floral scent, it had universal feminine appeal and it became the second best-selling perfume of all time, second only to *Chanel No.5*.

The fragrance, *Tabu (Dana),* was created by Jean Carles in 1932 with the instructions to create a perfume for a whore! He was at liberty to use a host of unusual materials with the resultant product an oriental, spicy-floral enchanting composition. It is composed of 10% patchouli oil with a dose of carnation. This perfume contrasted with Elizabeth Arden's launch of *Blue Grass* in 1934 with green fruit and floral notes and plenty of aldehydes.

On the men's side, the original *Old Spice* was launched in 1938. An Oriental accord perfume of warm spicy nutmeg, cinnamon and carnation with musk, benzoin and woody notes and a dash of *aldehydes* as well.

The 1940s

This was a decade with the outlooks divided in half, the initial WWII years of patriotism and shortages of food and clothing followed by the post-war years with a craving for "newness." Noted were knee-length A-line dresses with puffy shoulders and patriotic colors. The two-piece swimsuit was introduced and gender bending perfumes came on the market. The synthetics *hydroxycitronellal* and *musk ketone* were now more readily available to the perfumer (11,19,28-33,35,45,46,55).

Strides in perfume creativity came with the launch of *Bandit,* a bold concoction introduced in 1944 by Robert Piquet and composed by Germaine Cellier. It contained aldehydes and florals but what set it apart were the dark animal notes and leathery accord providing a classic *Chypre* perfume. The smoky, musky, leathery notes were created by incorporation of *isobutyl quinoline* and castoreum. The perfume, *Femme,* was also launched in 1944 with fruity top notes featuring plum, a spicy-flower heart and a dry down of sandal and musk created by the famous perfumer Edmond Roudnitska for Rochas.

Following WWII a profusion of excellent iconic perfumes came on the market. Jean Charles was the first to use *styrallyl acetate* in *Ma Griffe*, or *My Claw*, in 1946 which provided a green, fresh, floral aspect to the perfume. A huge amount of galbanum was incorporated into *Vent Verte*, created by Germaine Cellier also in 1947, to realize the green-floral genré.

Christian Dior's first foray into perfume with *Miss Dior* was with a debut catwalk show at the Avenue Montaigne store in 1947. There have been number "*Miss Dior*" type-perfumes reformulated in more recent years, but the original in 1947 was an avant-garde green, mainly Chypre floral perfume, with galbanum, Grasse rose, jasmine grandiflorum, Indonesian patchouli and Italian mandarin as naturals with synthetic *C10 & C11 aldehydes and styrallyl acetate,* balanced with the tarry, leathery castoreum.

In 1948 the sensational perfume, *Fracas,* was launched by Robert Piquet. This glamorous, sexy Parisian-style fragrance created by Germaine Cellier provided a profusion of white flowers. Notable are the seductive, exotic tones of tuberose and the sandalwood-musk base in the fragrance. *L'Air du Temps* by perfumer Françis Fabron for Nini Ricci was also launched in 1948 and was noted for the incorporation of a large volume of natural materials which resulted in an elegant composition.

Not to be ignored at the end of the 1940's was the introduction of *English Leather*, a popular notable fragrance for men. The appeal was a citrus with woodsy-leathery notes and an earthy, musky, smoky dry down, all mostly associated with the male gender.

The 1950s

The rise in prosperity and consumerism were apparent in this decade. Rock & Roll and the youth culture stepped out and a new age of celebrity with Elvis and Marilyn Monroe. Women's fashion featured soft blouses, slim sheath dresses and the poodle skirt. More pronounced animal scents inched their way into the compositions and *amyl salicylate* and *cedryl acetate* were exploited (11,19,28-33,35,45,46,55).

The introduction of *Youth Dew* in 1953 put Estee Lauder on the American fragrance map with an affordable long-lasting aroma. It was a modified bath oil that challenged the high-priced French perfumes on the market and has remained relevant into recent times. It is another aldehydic-floral with cinnamon notes and a diverse base of resin extracts and patchouli.

The perfumer Edmund Roudnitska went against the trend of sweet perfumes with the 1956 launch of *Diorissimo* for Chritian Dior. The perfume is inspired by lily-of-the-valley with additional ylang-ylang floral notes and a leafy green freshness. As noted previously, it is very difficult to obtain sufficient oil from the muget flower so it is probable that the composition contains some quantity of synthetic reproductions for lily-of-the-valley, such as *hydroxycitronellal*. Lily-of-the-valley (Muget) was to become the dominant theme of the 1950s.

L'Interdit is a perfume originally created for the elegant actress Audrey Hepburn by friend Hubert de Givenchy in 1957. It is a soft floral with creamy balsamic notes, a fresh, understated fragrance. *Vetiver*, also issued in 1957 (by Carven), was an early important masculine perfume. It has a very citrusy opening leading to a grassy vetiver and oakmoss-myrrh dry-down for a classic men's fragrance.

Rounding out the 1950s is *Cabochard* launched in 1959 by the fashion "Queen Gres." Created by perfumer Bernard Chant, it is a rose-jasmine composition with synthetic *ionones* for a violet note and *isobutyl quinoline* for the distinct leathery effects, combined with a harmony of smoky balsamic notes.

The 1960s

Youth and liberation were forefront in this decade. Flower power was the theme of the counter-culture, taboos were challenged, the "pill" was introduced and abortion legalized. Styles were also fragmented between the grunge of the "hippies" and the stylish mini-skirts. Every woman's closet had a pair of black trousers and a white top. There was also a wave of consumerism. A look towards the east was taking place in perfumery and scents were losing subtly. The sweet creamy floral synthetic *PTBCHA (4-tert-butylcyclohexyl acetate)* was now available for use in perfumery (11,19,28-33,35,45,46,55).

At the start of the decade, in 1960, Guy Robert created *Madame Rochas*, following on the floral-aldehyde theme. The green-citrusy opening and the classic floral heart, with some lily-of-the-valley, provides a fresh, uplifting fragrance.

The sixties started off women's liberation and with it came some liberating fragrances for men, *Brut (Faberge)* in 1964, *Aramis (Aramis)* and *Habit Rouge (Guerlain)* in 1965 and the oh-so special *Eau Savage* in 1966. The original *Brut* was a heavy musk Fougère while *Aramis* is a woody, strong leathery potion, a traditional, but still modern, perfume. Inspired by Jean-Paul Guerlain's Red Hunting Jacket, *Habit Rouge* is a woody Oriental fragrance, a take-off from Guerlain's *Shalimar* with an orange-opopanax twist.

Eau Savage by Dior soon followed on the market in 1966 created by perfumer Edmond Roudnitska. He dosed the juice with a slug (2%) of *hedione (methyl dihydrojasmonate)* imparting an impervious citrus floral-jasmine effect and with another synthetic, *helional*, for a fresh green floral aroma. It has a citrusy opening and musky male dry-down.

Not to be left waiting at the door, several great women's perfumes followed with *Fidji* in 1966 and *Calandre* and *Chamade* in 1969. *Fidji* was launched by Guy Laroche and has a distinct green leaf odor from galbanum and a floral bouquet with a notable scent of hyacinth. The success of Fidji influenced compositions into the next decade.

Calandre by Poco Rabanne starts off with the leafy green notes characteristic of the decade followed by florals with a strong rose aroma from incorporation of the synthetic *rose oxides* and ameliorated with jasmine. It is a sexy, seemingly metallic fragrance that includes the synthetic *Evernyl* (*Veramoss*) a natural isolate from oakmoss which carries the unique character.

Chamade is another masterpiece from Guerlain that evokes the wildness and loud notes of 60's women capturing the spirit of the decade. After a burst of bergamot, the heart is a unique combination of blackcurrant, hyacinth and the strong jasmine synthetic, *hedione*. The distinct freshness is softened by vetiver, orris and ambergris.

Chamade provides an ode to the 60s.

The 1970s

The energy crisis of 1973 was a sobering event but the decade was off to an energetic start with the controversial disco scene. The feminist, Germaine Greer, published her book "The Female Eunuch" giving rise to an almost universal feminist movement. Platform shoes contrasted with bell-bottom pants and corduroy pants & jackets. "Life-style" fragrances were introduced with the launch of some great blockbuster perfumes. The strong jasmine synthetic, *methyl dihydrojasminate* found extensive use during this decade (11,19,28-33,35,45,46,55).

The last fragrance launched by Coco Chanel was *Chanel 19* named for her birthday on August 19 when she turned 87 yrs old . It is another excellent perfume from Chanel composed by Henri Robert in 1970. The fragrance is a fresh, green floral with a touch of the throw-back scent of orris, resulting in a coolness compromised by the warmth of woods and leather in the base.

Aromatics Elixir was created for Clinique in 1971 by perfumer Bernard Chant who composed the perfume with a combination of unique twists that resulted in a woody-floral masterpiece. The opening aromatic notes of verbena, sage and chamomile moves to a combination of rose and white flowers, closing with oakmoss and patchouli. The fragrance engulfs and delights and is one of my personal favorites. *Rive Gauche* was also launched in 1971 by Yves Saint Laurent and is an exceptional floral-aldehyde. Green, rosy floral notes are combined with dark resinous effects to provide classical elegance.

Estée Lauder introduced *Alliage* in 1972 as a sports fragrance for women. The perfume is powdery green rather than sweet with an aura of chrysanthemums. The pine and rosewood middle notes move on to the vetiver, myrrh and musk dry-down. A new combination for the 70s woman.

Coriandre (Jean Couturier) launched in 1973 was never a huge seller but had a strong following for the unique character of this fragrance. The inclusion of angelica in combination with coriander provided a herbaceous, fresh vibrancy. This opening is followed with a swatch of synthetic rose *damascones* with orange blossom and geranium, giving way to a definite musky, civet darkness.

Naming of the perfume *Opium* by Yves Saint Laurent in 1977 was certainly a declaration of a level of decadent hedonism for this fragrance. The design of the original bottle chosen for the perfume was based on a small case used by Japanese samurai for storing their medicines and opium. The Oriental themed perfume is a spicy, exotic blend of lush flowers and deep wood notes. An exceptional fragrance and a huge success on the market.

The year 1978 witnessed the launch of three more great fragrances of *Anais Anais*, *White Linen (Lauder)* and the men's perfume *Polo (Ralph Lauren)*. *Anais Anais* by Cacharel was produced by a "gang of four" perfumers, which is rather unusual, but their combined expertise created a product which returned the market to a more sweet, ultra-feminine fragrance after the success of the distinctly different *Opium*. The bottle of a pink floral cover reflected the character of *Anais Anais* which was indeed a quality product, but very affordable. It was composed of a bevy of classic florals, from the top of orange blossom and hyacinth to the heart of rose, jasmine, tuberose and lily-of-the-valley, and ending with the warmth of sandalwood with amber.

Of all the named perfumes, *White Linen* to me, most closely reflects the expectations one perceives from the label. The perfumer Sophia Grojsman created the composition with a sharp, clean, healthy appeal. The effect is undoubtedly due in part to the soapy *aldehydes* but they are both tempered and enhanced exquisitely with notes of rose and jasmine, a middle note of orris leading to a vetiver/oakmoss base.

The third notable perfume of the 1978, *Polo,* is a masterpiece for men introduced by Ralph Lauren. Chamomile is employed for a grassy but slightly floral note combined with herbs of basil and thyme and spices of cumin, coriander and cloves. But the smoky leather and tobacco provide that kick for the masculine appeal.

The 1980s

The digital revolution has emerged to never be slowed down. "Yuppies" personified the new capitalism. Women's shoulders were rounded with shoulder pads and leggings were the fashion, one of the most eclectic decades in fashion. Perfumes evolved to be more strong and loud. The floral watery herbaceous *helional* and sweet musky *ethylene brassylate* came into common use in perfumes (11,19,28-33,35,45,46,55).

In 1985 two great perfumes were launched as *Poison (Dior)* and *Beautiful (Lauder)*. Some say *Poison* defined the 80's and it certainly was as shocking a name as the previous *Opium* in the 70's. The perfume reinforced Dior as a premier house of exceptional fragrances. The perfumer, Edouard Flèchier, used plenty of both fruit and floral essences in the composition, black & red currants, blackberries and raspberries with rose *damascenones*, jasmine and a huge tuberose note. All tempered with a woody-spicy end featuring opopanax. The elegant perfume won a FiFi award from the Fragrance Foundation in 1987 for creative achievement. *Poison* was followed by *Beautiful* which continued the sweetness of the decade. The perfume was a lustful bouquet of florals with aromatic green and amber notes.

Calyx is an outstanding perfume launched in 1986 by Prescriptives. The perfumer

Sophia Grojsman topped out the potion with grapefruit and then loaded it with tropical fruit essences such as guava, mango, papaya and passion fruit along with plenty of exotic florals including rose, freesia, marigold and cyclamen. The truly innovative additive was the aldehyde, *helional*, the green, watery, marine scent utilized previously in *Eau Savage* and *Diorella*. The result was a sweet, green, watery floral, setting the stage for the "oceanic" scents of the future.

Two men's fragrances were introduced in 1988, *Fahrenheit* by Christian Dior and *Cool Water* by Davidoff. Both fragrances saw substantial success. The lavender and nutmeg in *Fahrenheit* immediately suggest a more masculine effect compounded with the strong scent of violet in the heart from the synthetic *methyl heptin carbonate*. The close with the notable leather, vetiver and patchouli notes summarizes this encompassing men's fragrance.

Cool Water was a definite trend setter for many decades to follow. Perfumer Pierre Bourdon started off the fragrance with minty, lavender green notes and a flower heart but the innovative use of two synthetics really set the perfume apart from all the previous masculine fragrances. *Dihydroxymyrcenol* provided a soapy like freshness while the other synthetic, *Calone*, offered up the ocean as a characteristic sea water scent. Never fully matched by subsequent numerous attempts.

Samsara came on the scene near the end of the decade in 1989 with a powerful scent created by Guerlain. The fragrance is sweet, modern and sophisticated. It has fruity-citrusy top notes and jasmine, iris, violet and rose in the heart, but what sets it off is the woodsy effect of sandalwood, both as a natural of Indian origin and the synthetic sandalwood note, *polysantol or mysantol*. These synthetics diffuse the characteristic creamy, powdery sandalwood aroma through all levels of the perfume, from top to base.

The 1990s

The decade began with a global recession and a move away from consumerism to minimalism. Social conscience grows stronger, taboos about homosexuality are constrained and environmentalism comes to the forefront. A New Age of spiritualism came out of California. Halter-neck and tube tops emerged with tie-dyed shirts adorning the youth. It was in this decade that the bare midriff really took off from celebrities to teeny boppers.

A return to nature and a strong theme of escapism is in the air and fresh new perfumes with Aquatic/Oceanic accords are embraced. The fresh citrus floral synthetic, *dihydromyrcenol* was exploited and additional synthetic musks were added to the perfumer's palette during this decade (11,19,28-33,35,45,46,55).

At the start of the decade in 1990, the perfume, *Trésor*, was launched by Lancôme.

The perfume was created by Sophia Grojsman based on her innovative "horizontal" approach to perfume composition described in a later section. The fragrance was a great success even though it was a rather heavy, sweet potion. The structure combines natural florals with a her classic potion of a whole lot of synthetics such as *methyl dihydrojasmonate* with a strong citrus jasmine effect, *Ionone* for a floral-violet scent, *iso E Super* providing a woody amber note and, *Galaxolide* a clean musky floral contribution.

Amerige (Givenchy) launched in 1991 is probably one of the last really sweet hold-outs from the previous decade and can be overpowering. It has plenty of fruity, floral notes and is additionally dosed with 30% *methyl dihydrojasmonate,* that strong jasmine derivative, 25% of *benzyl salicylate* providing a faint sweet floral balsamic odor and *iso E Super* for the woody amber note. The result is potion adored by some and despised by others.

Escape was also introduced in 1991 by Calvin Klein. It was notable as an early aquatic-type scent taking advantage of the synthetic *calone* which provided a watery world with melon overtones. The heart contained a package of sheer florals of cyclamen, lily-of-the-valley, rose, heliotrope and carnation rounded out with a base of woods and musk.

In 1992 the truly unique perfume, *Angel*, was introduced to the market by Thierry Mugler. The star shaped bottle was also very distinctive. Mugler convinced the perfumers Oliver Crisp and Yves de Chiris to create a perfume that reminded her of her childhood trips to the fairgrounds with the smells of cotton candy and toffee apples. They achieved this with incorporation of a new synthetic, *ethyl maltol*, which has a caramel aroma, as well as honey essence. The opening of bergamot and *methyl dihydrojasmonate* was followed by the fruity sweetness of strawberry, coconut and the synthetic *neocaspirene*, a black currant bud scent, all for a tropical "fruitchouli" aroma. Several classic florals were included and a noticeable strong patchouli base, which also created a chocolate aroma in combination with the *ethyl maltol*. This exceptional new "Gourmand" fragrance was an immediate hit, with many other houses attempting to duplicate the effect. The perfume won a FiFi Hall of Fame Award from the Fragrance Foundation in 2007 for creative achievement.

CK One (Calvin Klein) was a pioneer "unisex" perfume in 1994 and also sometimes referred to as a "skin scent." The fragrance was inexpensive and affordable and became a best seller. It was composed of huge amounts of synthetics, citrus fresh from the start, with a host of florals to follow and a large amount of synthetic musk for the base.

Following on the gourmand chocolate theme of *Angel*, the perfume *A*Men*, a "chocolate mint" type fragrance was designed for men in 1996, but also equally suitable for women. The perfume starts off with metallic mint-lavender notes and some *aldehyd*e sparkle combined with some heavy woody notes and a kick of caramel, coffee and styrax. The fragrance has been an inspiration to many that followed.

Acqua di Gio nicely followed the 90's trend with a fresh, clean, watery fragrance introduced by Giorgio Armani in 1996. The perfumer Alberto Morillas incorporated the synthetic *calone*, the so-called "watermelon ketone," in the potion to also provide the aqueous freshness similar to previous oceanic type fragrances. It is a woody-spicy fragrance with aromatic notes of rosemary, sage and geranium and an incense-patchouli base.

Happy by Clinique became a super best seller after the launch in 1997. Created by perfumers Jean Claude Delville and Rodrigo Flores-Roux with a combination of bergamot, apple, and plums for the opening with a special floral heart of Mandarin blossoms, laurel, orchid and rose. This was all followed with a unique base of mimosa, lily, magnolia, amber and musk. The popular perfume won a FiFi award from the Fragrance Foundation in 1998 for creative achievement and it has inspired many imitators.

Closing out the 90's was the launch of *Lime, Basil & Mandarin* for both men and women by Jo Malone. It is an enlightening fragrance with the crisp citrus notes and the herbal and floral accompaniments of basil, thyme, lilac and iris. The dry-down is to a grassy vetiver-patchouli aroma.

The 2000s

The new millennium opened with a strong economy and enormous optimism. The society moved on from the "9/11 disaster" and became the decade of communication, with the mobile phone ubiquitous. Hold-overs from previous decades, the track suit, hip-huggers and tank-tops continued as a fashion statement of casualness for the society. However, the perfume industry was rapidly being taken over by huge corporations with concomitant reductions in the quality of the fragrances. Celebrity perfumes were omnipresent and a separate section is devoted to this class of perfumes (11,19,28-33,35,45,46,55).

Thousands of perfumes started appearing on the market approaching the new millennium which, of course, provides numerous options for the consumer. Many of the perfumes were not of the highest quality with minimal ingredients and relying mainly on synthetics. The macrocylic musks were replacing the previous other synthetic musks in formulations. A few selections are described for this period.

In the first year of the decade, in 2000, the perfume *Flower* was introduced by Kenzo. It is an Oriental Floral with powdery, sensual, soft notes of violet, hawthorne, cassis and Bulgarian rose. Several notable synthetics also appear in the formulation as *hedione* for the fresh, floral muget notes and *heliotropin (methyl dihyrdojasmonate)* for warm floral jasmine aroma. It is said to be a knock-off of the 1923 fragrance, *Royal Bain de Champagne,* by Caron (45,46).

Coco Mademoiselle was launched in 2001 by Chanel. The perfumer Jacques Polge combined notes of orange and lychee with patchouli to create this flirtatious fragrance. This fruit-patchouli combination, sometimes referred to as "fruitchuli," was utilized for the creation of the blockbuster perfume, *Angel*. The outstanding success of *Coco Mademoiselle* was unexpected since it was not a featured perfume by Chanel. It also received a FiFi Award from the Fragrance Foundation for the Best National Advertising Campaign on TV in 2008.

The name alone for this 2005 launch of *Flowerbomb (Viktor&Rolf)* signals where your scents are about to step. It provides an explosion of a bouquet of flowers, orchid, sambac jasmine, Provence rose and freesia. Bergamot and tea provide a fresh opening and the fragrance closes with a patchouli and musk base. This Oriental floral perfume received a FiFi Award from the Fragrance Foundation for the Best National Advertising Campaign for print in 2006.

Euphoria was also launched in 2005 by Calvin Klein and also follows a theme similar to *Angel* with the fruity-patchouli (fruitchouli) combo, with addition of some cotton candy aroma. It is a complex perfume created by perfumers Dominique Ropion, Carlos Benaim and Loc Dong. They were able to distinguish the potion from the predecessors with the right combination of dark exotic flowers, fruit notes of pomegranate, raspberry, passionfruit, persimmon and peach and base notes of mahogany, amber, musk and, of course, patchouli. It also won a FiFi Award from the Fragrance Foundation in 2006.

Tom Ford followed with *Black Orchid* in 2006. A sensual luxurious fragrance that opens with a unique combination of black truffle, black currant and a mélange of citrus. The orchid and spices with ylang-ylang and fruity notes in the heart are followed by characteristic base notes of the decade of patchouli and dark chocolate. The complex beauty of the fragrance allures both women and men.

Very Sexy introduced in 2007 by Victoria Secret was created by perfumer Jean Claude Delville. It is a fresh creation with, what some say, a metallic aroma. The top is not composed of the usual notes, but of pepper, cactus flower, cappuccino with a touch of clementine and the heart has camellia, mimosa, hydrangea and orchid. The dry-down imparts a woodsy white amber combined with a scent of blackberry.

The bit over-the-top fragrance, *Viva la Juicy*, was launched in 2008 by Juicy Couture. It fits well as a Fruity Floral Gourmand, a playful flirty fragrance for women. Wild berries and mandarin are combined with honeysuckle, gardenia and jasmine leading to the delicious base of caramel, vanilla, praline, amber and sandalwood. The bottle features a gold royal insignia on the front wrapped with an adorable red bow.

The 2010s

With the continued rise of Facebook, You-Tube, Twitter, Instagram, etc., it seemed that everything was "Going Viral." Hip-hop and R&B music reigned and divisions in society were promoted with the concept of "Fake News." Women were layering shorts over leggings and bomber jackets took off again. The Gourmand Fruity fragrances were continuing to climb in popularity influencing approaches to perfumery (33,46).

Acqua di Gioia is a Floral Aquatic fragrance launched at the start of the decade in 2010 by Giorgio Armani. Armani treasured his summer holidays by the sea at his island villas and he wanted a fragrance that captured the seaside. Aromatic mint and lemon are first realized from the potion and likely the presence of the synthetic, *calone*, known to evoke an oceanic appeal. The heart is peppery, peony and jasmine and finally a dry-down to cedar, brown sugar and amber-labdanum.

Following in the theme of the successful *Viva la Juicy, Prada Candy* was introduced in 2011 by Prada and composed by the perfumer Daniela Roche Andrier. The perfume has strong doses of quality materials providing a special sensual experience. Neither florals nor citrus are featured in this composition, but rather you are immediately drawn in by the tasty aroma of caramel with powdery notes. The base of vanilla and balsamic benzoin holds your attention to this novel olfactory fragrance.

La Vie est Belle for "Life is Beautiful" is an opulent, joyful but elegant perfume launched by Lancome in 2012. It has been reported that the three perfumers responsible for the fragrance, Oliver Polge, Anne Filpo and Dominique Ropion, worked with the house for three years and 5000 versions to capture the final product. It is a Floral, Fruity, Gourmand fragrance with a start off of juicy black currant and mellow pear, followed by a heart keyed with iris with jasmine and neroli and a gourmand base of sweet praline, warm vanilla and Tonka Bean. Although actress Julia Roberts was the face of the perfume, it is not considered a "celebrity perfume" which are described in a following section.

A*Men Pure Leather was also launched in 2012 by Thierry Mugler as part of a new limited edition of *"Les Parfums de Cuir"* or the Fragrance of Leather.
The fragrance has a sweet, spicy and certainly leathery scent. The top notes already emanate leather, mint and fruity notes combined with coriander and green notes. Leather again in heart with caramel, jasmine and honey. Of course, the base note is leather with tonka bean, sandalwood, coffee and vanilla. It has been reported that the leather aroma is actually derived from soaking natural leather in the original composition for four weeks, harking back to the early history of perfumery.

Oud Oud Oud (Stephanie de Bruijn) appeared on the market in 2013 composed by perfumer Stephanie de Bruijn. This is a mysterious powerful Oriental fragrance eliciting the charms of the Middle East. Turkish rose and rosebay-willow-herb are in the top notes followed by a strong shot of agarwood (oud) with patchouli, benzoin and cedar in the heart and softened in the base with Madagascar vanilla, sandalwood and leather. A magic unique contribution.

My Burberry was issued by Burberry in 2014 as a floral fragrance for women. The perfumer Francis Kurkdjian related that the fragrance was inspired by an urban London garden after the rain; clouds, rain and flowers. The fragrance starts off with sweet pea and bergamot, with middle notes of peach, gardenia and geranium with a minty herbal contribution. The fruity floral heart also contains quince providing an odor between pear and apple. The base is a unique combination of several rose scents with patchouli, musk and leather.

Wonderlust was launched by Michael Kors in 2016 and promoted with the idea of feminine escapism. An Oriental Floral for the modern feminine woman with an infinite desire for discovery. Uniquely almond milk, pink pepper and bergamot form the top note, with heliotrope, jasmine Sambac and carnation in the heart and Cashmere wood, benzoin and sandalwood in the base. It is a sweet, spicy, creamy floral with and an almost edible appeal.

Launched in 2017, Mugler's *Aura* immediately draws your attention with the deep green fragrance beaming out in the heart shaped container. Apropos to this image, the perfume is a Green Oriental comprised of slightly biting rhubarb leaf and bergamot at the top, with green notes, pear and ylang-ylang in the middle tier and woods, amber, coumarin and vanilla, with a touch of bourbon for the base. One of the most talked about launches of the decade.

Nomade issued in 2018 is another perfume with the theme of feminine escapism, the essence of a free-spirited and confident woman. The Floral Chypre composition was created by the perfumer Quentin Bisch for Chloé. The top note is dominated by a fantasy fruity accord of Mirabelle plum with a sweet, fresh plum aroma combined with citrus. The bright floral aroma of the heart is carried by the freshness of freesia accompanied by peach and rose. A dominate oakmoss with sandalwood nicely closes out the experience.

The 2020s

Not quite the roaring 20s of the 1900s, the appearance of Covid-19 in 2020 set the world on its back. A new American president and a more conciliatory mode prevailed with some interesting new fragrances appearing on market, a few introduced below (11).

Named for the American fashion designer herself, *Rebecca Minkoff* was launched by Rebecca Minkoff early in 2020. The sensual, warm Oriental Floral fragrance was composed by Linda Song and Rodrigo Flores-Roux. It composed of a modern tobacco accord with the feminine effects of jasmine and the sweet herbaceous aroma of the synthetic, *amyl salicylate*. The fragrance has a warm magnetic aura.

Voce Viva was launched by Valentino and composed by two women perfumers, Amandine Clerc-Marie and Honorine Blanc. The Floral Musk Woody fragrance is an ode to the voice of femininity, intimate and personal. The ginger citrus top note with an unexpected cool affect feeds the white flower heart and closes with sweet pleasant notes of vanilla, tonka bean and sandalwood.

The Oriental Floral fragrance *Lil Fleur* for both men and women was launched in June, 2020 by Byredo. The inspiration for the perfume is youthful intensity, boldness and openness. A crisp sparkle is provided by the top notes of cassis and tangerine evened out with soothing saffron, an enrapturing heart is provided with Damask rose and leather and a smooth base of vanilla, ambergris and woody notes.

A break from the routine was provided with the launch of *Bitter Peach* by Tom Ford in October, 2020. This exotic contribution has a complex composition. Peach, blood orange and cardamom in the top notes, with a visit to the liquor cabinet in the heart with rum and cognac, tempered with the sweet, tea-like aroma of the aromatic herb, davana, all combined with sensual jasmine. The woody, resinous undertone is realized with a variety of base notes including styrax, benzoin, labdanum, vetiver, sandalwood and patchouli.

The fragrance *SOL Cheirosa '62* was composed by perfumer Jerome Epinette and launched by Sol de Janerio. Cheirosa translates to "smelling" and for every Brazilian it is for smelling delicious. The Oriental Vanilla fragrance captivates the sounds and sensations of summer with sweet pistachio and almond top notes, a floral heart and additive gourmand notes of caramel, vanilla and salt, with sandalwood in the base.

Bond No 9 launched the Oriental Floral fragrance *My New York* for both men and women in 2020. Based on the international phenomenon of what is New York City, the idea of this fragrance is the global, diverse, vibrant and inclusive character of the city reflected by a world of flags on the distinct packaging. The sexy, spicy perfume opens with a fresh piney spicy odor from elemi resin combined with pink pepper and ginger. The rose floral heart is complimented with the dry aromatic woody aroma of papyrus and the warm woody musky amber scent of cashmeran with sandalwood and patchouli rounding out the base.

Celebrity Perfumes

While the images and endorsement of celebrities have been utilized for promotion of perfumes over the past 100 years, it was not until 1981 that the first celebrity perfume, as we know it today, was launched as *Sophia* for Sophia Loren. There are now over 500 celebrity fragrances on the market representing a list of 75 different stars from film to sports. The sales, though diminishing in recent years, have been in the billions of dollars and clearly the fragrances have been a huge marketing success. A few special ones are described as follows (11,19,28-33,35,45,46,55).

This first blockbuster perfume in this category was *White Diamonds* for Elizabeth Taylor in 1991. My mother loved this perfume and it has racked up sales of over $77 billion. It is a sheer floral, soft woody fragrance created by perfumer Carlos Benaim with top notes of aldehydes and citrus, heart notes with a classic trio of rose, jasmine and violet, a tropical contribution of tuberose and ylang-ylang and a base containing oakmoss, patchouli and sandalwood. The perfume won the FiFi award of "Fragrance of the Year" and was inducted into the Fragrance Hall of Fame in 2009. *White Diamonds* remains one of the best-selling celebrity fragrances in the world. The success of Elizabeth Taylor fragrances, now numbering eighteen, was a prime motivator for celebrities to launch their own signature perfumes.

Glow was launched in 2002 by singer and actor Jenifer Lopez. The target group for the fragrance was young women, late teens to early 20s. *Glow* rapidly became a best seller, contributing $300 million in the first year, and spawned 30 more fragrances from her label. It is a clean, sweet, floral with powdery notes and a "silver spoon" effect provided by the synthetic *helional*. J.Lo described the fragrance as a skin scent, as though you just came out of the shower and are the sexist person in the world! Citrus top notes are followed by the classic rose, jasmine and tuberose, drying-down to musk, sandalwood, vanilla and amber.

In 2005 *Lovely* was launched for Sarah Jessica Parker, composed by perfumers Clement Gavarry and Laurent Le Guernec with Coty. It is a charming soft, powdery, discrete and very intimate fragrance, a silky white amber. It possesses an elegance characteristic of classic fragrances with aromas of lily-of-valley and magnolia. Chandler Burr, the former scent critic for the New York Times, has described the development of this perfume in his book "The Perfect Scent" published in 2007 (47). Sarah Jessica Parker actively participated in the creation of the perfume, which was rather atypical, and features of the creation of the fragrance are described in a following section.

The singer performer Britney Spears issued her first perfume, *Curious,* in 2004 to instant success of $100 million in the first year. The fragrance was followed by 30 more releases, with her recent *Glitter Fantasy* in 2020. *Curious* is a fruity floral accented with magnolia and cherries wrapped in a sweet vanilla-infused musk.

Some critics call it a bit syrupy but her later launch of *Midnight Fantasy* in 2006 received better acclaim with its exotic fruits and night flowers. It is still a fruity floral with top notes of plum, cherry and raspberry, a heart of orchid, iris and freesia and base notes of vanilla, musk and amber. The Britney Spears line of perfumes have brought in more than $1billion in sales for Elizabeth Arden.

Michael Jordan, launched by Michael Jordan the star basketball player, was launched in 1996 as a "*Young Spice*" for men. The name is a take-off from "Old Spice," the spicy cinnamon-carnation men's popular fragrance. *Young Spice* is cool, crisp citrusy with notes of lavender, juniper and fir with an undertone of wood and musk. The first year's sales were good, $9 million, followed by $5 million in year two and then dropped off. Jordan has issued a total of 12 fragrances (five for men) with the most recent in 2014. A host of additional men's fragrances have been issued honoring sport's heroes.

Global Scale of Perfumery

The Global Sales of fragrances are estimated at $33 Billion per annum and is expected to grow at a rate of near 4% from 2020 to 2025. This growth is due to the increasing interest in personal care and the demand for exotic and luxury fragrances. The rise in income status and improved living conditions are enhancing the global market for perfumes, cosmetics and personal care products. Also, the enhanced creativity of perfumers and product diversification are expected to expand the customer base. Fine Fragrances represent about 15% of the global fragrance market with the rest of the sales for soaps & detergents (~30%), cosmetics & toiletries (~30%), household products (~20%) and others (~5%) (15,34).

The total fragrance industry sales in the USA are about $6 billion annually. Four out of five women in the USA state that the use a fragrance product on a regular basis. Although women purchase the majority of perfume fragrances, men are garnering an increasing share of the market. Women tend to prefer on-site evaluation of perfumes but online sales have steadily increased and are now at about 30%. Online purchases are expected to increase substantially in future years.

There are hundreds of new fragrances launched each year and there are over 1,000 fragrance brands to choose from in department stores. Surveys have found that millennials tend to favor perfumes created from natural ingredients and sales have moved away from mass market products to more personalized fragrance formulations.

The fragrance industry is quite consolidated into a number of major corporations and many known perfume lines have been subsumed by the industry giants. The major corporations and their approximate market share are shown in Table 2.4. The percentages include the market share for both fragrances and flavors (34,55).

The five major Conglomerates listed in Table 2.4 account for about 67% of the global sales of fragrances and flavors. Coty holds a major share of the perfume market but there are other independent players in the market. The market is dynamic in terms of take overs, i.e., Frutarom was recently acquired by IFF, so some of the companies in the following list may become part of the conglomerates in the future: Sensient, Robertet, Hasegawa, Huabao, LMR, LVMH, L'Oréal, Payan Bertrand, Estée Lauder, Avon, Chanel, Shiseido, Beiersdorf, etc. (11,34,55,57).

Table 2.4 Global Market Share of Flavors & Fragrances for Major Corporations

Corporation	Headquarters	Market Share, %
Givaudan	Switzerland	20%
Firmenich	Switzerland	19%
Int'nl Flavor & Fragrances	New York	13%
Symrise	Germany	10%
Takasago	Japan	5%
Mane SA	France	5%

Estimates from 2017 except Fermenich 2020. References: Leffingwell (34), Pybus (55) & Teixeira (57).

Chapter 3. Creation of a Perfume

Perfumes are composed by a perfumer or a "nose" who usually has years of extensive training to hone his or her olfactory skills. All of the great perfumers expound on the importance of developing a solid knowledge of hundreds of odors, at least 200-400 with extra exceptional familiarity. Great insight into the creation of perfumes can be realized by analyzing the olfactory training methods utilized to obtain sufficient knowledge of both natural and synthetic odors as follows.

Olfactory Training Methods

Two great French perfumers, Jean Carles and Edmond Roudnitska, both once affiliated with the prestigious Grasse Institute of Perfumery, have described their methods for training of the nose for perfumery.

Jean Carles (38,39) stated that most students of perfumery can become expert with sufficient dedication and training. The perfumer works with dozens of natural essences (essential oils, absolutes, etc.) and thousands of chemicals. In the past the commitment to training amounted to over ten years of study. The French perfumer, Jean-Claude Ellena stated that "It took ten years to know and twenty to master" the art of perfumery (58,59). Still many months to years need to be apportioned to attain sufficient expertise. An International Technical Degree in Fragrance Creation and Sensory Evaluation from the Perfumery School in Grasse requires an 18-month commitment which includes a six-month internship, while training as a Technical Assistant in Fragrance, Flavor and Cosmetics at the school takes six months.

Jean Carles utilized a method of similarity and contrast for olfactory training. He used a series of olfactory charts for odor analysis of both natural [150] and synthetic [350] raw materials. Sample hypothetical charts, such as those utilized by Carles and similar forms developed by many perfumers since, are shown in Tables 3.1 and 3.2 (38,39,60). The descriptions in the tables, based on literature references, provide a good idea of the approach for training perfumers to the odors of natural and synthetic materials. By moving vertically down the tables starting at the top 1st study, the student learns by contrast about the very different odors from the various families of raw materials. Then by moving horizontally across the charts the student can compare the variety of odors from each different family. The method of study has the advantage of not tiring the nose as would arise with studies of only similar materials. This "Olfactive Study by Contrast" remains as the basis for teaching perfumery students today (38,39,60-62).

Table 3.1 Olfactory Study of Natural Perfume Raw Materials

		Study of Contrasting Odors*					
Study	Notes	Study 1	Study 2	Study 3	Study 4	Study 5	Study 6
7th	Citrus	Lemon	Bergamot	Mandarin	Orange	Lime	Grapefruit
8th	Woody	Sandalwood	Cedarwood	Vetiver	Patchouli	Oakmoss	Guaiacwood
9th	Spicy	Clove	Cinnamon	Nutmeg	Pepper	Juniper	Coriander
10th	Orange	Neroli oil	Petitgrain	Orange^			
11th	Anise	Anise	Basil	Fennel	Tarragon	Caraway	
12th	Rose	Rose^	Rose oil	Geranium^B	Geranium^G	Geranium^P	
13th	Rustic	Lavender	Lavendin	Rosemary	Eucalyptus	Hyssop	Clary Sage
14th	Amber	Cistus^					
15th	Balsamic	Peru Balsam	Tolu Balsam	Vanilla	Tonka Beans	Styrax	Cistus
16th	Floral	Jasmine	Tuberose	Ylang-Ylang	Mimosa	Orris^	Muget
17th	Fruity	Blackcurrant	Osmanthus^				
18th	Resin	Olibanum	Benzoin	Opponax	Myrrh	Elemi	Galbanum
19t	Green	Galbanum	Violet leaf				
20th	Citral	Lemongrass	Verbena	Melissa	Citronella		
21st	Minty	Peppermint	Spearmint	Marjoram			
22nd	Animalic	Civet	Musk	Castoreum	Ambrette(seeds)		

* Hypothetical sample model of Jean Carles' Olfactory Training Method (oils unless specified), ^{B}Bulgarian, ^{G}Grasse, ^{P}Palmarosa, ^Absolute; References: Carles (38, 39), Kydd (60)

Table 3.2 Olfactory Study of Synthetic Perfume Raw Materials

Study Notes		Study 1	Study 2	Study 3	Study 4	Study 5
		\multicolumn{5}{c}{Study of Contrasting Odors*}				
6th	Citrus	Citral	Citronellal	Dihydro-myrcenol	Grapefruit acetal	
7th	Anisic	Anisic aldehyde	Anethole			
8th	Aldehydic	Ald. C8	Ald. C9	Ald. C10	Ald. C11	Ald. C12 (MNA)
9th	Balsamic	Cinnamyl acetate	Cinnamic alcohol	Amyl salicylate	Isobutyl salicylate	
10th	Woody	Iso E Super	Vetivenyl acetate	Veramoss	Sandalore	Javanol
11th	Amber	Ambrofix	Kephalis	Cedrylmethylether		
12th	Green	Phenylpropyl Alcohol	Hexenol-3-cis	Gardenol	Methyheptine carbonate	
13th	Marine	Calone	Azurone			
14th	Spicy	Cinnamic ald.	Eugenol	Isoeugenol	Methyleugenol	
15th	Floral	Linalool	Hydrocitronellal	Terpineo	Isoraldeine	β-Ionone
16th	Rose	Phenyethyl Alcohol	Geraniol	Citronellyl acetate	β-Damascone	Rose Oxide
17th	Jasmine	Benzyl acetate	Hedione	Benzyl salicylate	Benzyl propionate	γ-Jasmine lactone
18th	Orange	Methyl Anthranilate	Linalyl acetate	Argeol	Linalyl benzoate	
19th	Fruity	γ-undecalactone	Amyl acetate	Hexenyl acetate	Raspberry ketone	Amyl benzoate
20th	Sweet	Vanillin	Heliotropine	Coumarin	Ethyl maltol	
21st	Musky	Musk Ketone	Galaxolide	Helvetolide	Nirvanolide	
22nd	Leathery	Isobutylquinoline				
23rd	Animalic	Indole	Skatole			

* Hypothetical sample model of Jean Carles' Olfactory Training Method.
References: Carles (38, 39), Kydd (60)

Roudnitska (63-66) has described some of the features of his training program where he emphasized the main elements of olfactory perception of intensity, duration and volatility all of which contribute to quality of the odor. Memory of the odors and, indeed, of the many complex olfactory combinations, are necessary to master an extensive olfactory palette. Roudnitska's approach to memorizing the various odors was to group them in families or series not unlike Carles' method. He utilized 15 different series with each series consisting of a selection of natural products and their main chemical constituents. The groupings were: Citrus, Rose, Orange, Jamine, Violet-Orris, Aniseed, Aromatic, Green, Spicy, Woody, Tobacco, Fruity, Balsamic, Anamalic and Leathery.

The attributes to be noted by the student in the study of each of the series include quality and character of the odor, intensity, expansion mode (diffusion/volume), stability, evolution of note and duration of perceptibility. The students record everything that comes to mind with words that arise spontaneously. They are requested to ask questions such as "Do the notes create an image or enable a thought to be more precise?" Roudnitska emphasized that it is the first impression that is the purest interpretation of the odor. He employed smelling strips for his program (63-66).

Roudnitska revealed his approach for the study of Citrus as a series of sequential groups of odors for student evaluation (at 10% in ethanol unless otherwise specified), starting with (63-66):
Group I (comparison of essential oils)- lemon, bergamot, orange and tangerine.
Group II (Bergamot & constituents) - bergamot oil, linalool, linalyl acetate, terpineol, terpenyl acetate, limonene.
Group III (Lemon & constituents) - lemon oil, lime oil (linden), verbena or lemon grass oil, citral (1%), limonene.
Group IV (Orange odors) - Guinea orange oil, Florida orange oil, Bitter orange oil, C-10 aldehyde (decanal, 0.1%), methyl anthranilate (1%), limonene.
Group V (Tangerine odors) – tangerine oil, C-10 aldehyde (decanal, 0.1%), methyl anthranilate (1%), dimethyl anthranilate (1%), limonene.

Roudnitska then provided a series of exercises testing the student's skill in identification of constituents and their proportions in simple blends of, for example, a Neroli oil (six constituents), an Orange flower concrete with a few ingredients and a mixture with the note of Bergamot.

The training program for one additional grouping, Rose, of Roudnitska's 15 different series is shown below to demonstrate the sequence of sampling and necessary breaks to adjust the nose. The list becomes increasingly complex with chemical constituents of rose oil. Concentrations are 10% in ethanol unless otherwise specified in parentheses (63-66).

Rose Odorants
Morning session I (1hr)
Rose oil (Bulgarian then Moroccan)
Rose de Mai absolute (Grasse)
Rose absolute (Moroccan)
Geranium oil (Bourbon then Moroccan)
Palmarosa oil
[Five minute break outdoors]

Morning session II (1hr, 45 min)
Smelling first the alcohol compound then the corresponding esters:
Geraniol then Geranyl acetate, then Geranyl propionate
L-Citronellol then isomer D-Citronellol, then Citronellyl acetate, then Citronellal, Rhodinol, then Rhodinol acetate
Nerol then Nerol acetate

[Fifteen minute break outdoors]

Morning session III (1 hr)
Linalool then Linalyl acetate
Phenyethylalcohol
Cinnamyl alcohol then Ethyl cinnamate
[Five minute break outdoors]

Morning session IV (45 min)
Eugenol
n-Nonanol (1%) then n-Nonanal (0.1%)
Styrallyl acetate

Additional characteristics for categorizing each new odor is a scale of olfactory sensations from cool-warm, light-heavy, dry-fatty, bitter-sweet, acid-smooth, sharp-deep, ethereal-balsamic, flowery-fruity, green-candy, and transient-tenacious. But note that each odor can have designations in several categories, i.e., Lemon (sharp & transient); Amyl acetate (ethereal, fruity & transient); Labdanum (deep, balsamic, tenacious & fruity); and Violet absolute (green, fatty, tenacious & flowery). These benchmarks can be used for the choice of a material for a composition (63-66).

As previously discussed, Jean Carles (38,39) established the olfactory fragrance pyramid shown in Figure 2.2 for use in composing vertical fragrances. His methodology is the classic approach for composing both accords and perfumes based on volatility of the odorous materials. The volatility of fragrance materials is based on scientific analysis, but a student can ascertain the tenacity by noting the time it takes for a smelling strip to lose the characteristic odor. As shown in Table 2.1, the top notes are very volatile and lack tenacity, while the middle notes have intermediate volatility and tenacity and are also referred to as modifiers and the base notes have the lowest volatility and highest tenacity. This volatility classification shown in Table 2.1 is for just a small fraction of the possible notes available to the perfumer.

The base notes are an important component of the perfume since this aroma can last for many hours. However base notes such as vetiver, oak moss and patchouli in the Chypre accord can be rather unpleasant when initially smelled, but over time, with evaporation, they take on a very pleasing odor. Modifiers of intermediate volatility (middle notes) are added to alter the initial unpleasant effect of the base notes. The base notes also provide fixative properties enhancing the longevity of the fragrance. The top notes of high volatility are then added to impart a very pleasant odor upon opening of the bottle of perfume.

The next step is creating accords from notes, that is, creating distinct aromas from a composition, usually of 3-6 different essential oils, notes and synthetics. The construction and study of a Chypre accord is shown below to demonstrate Jean Carles' training methodology (38,39). However, a number of the notes are no longer available so some substitutions are suggested. As a reminder, typical base notes of a Chypre accord can include oakmoss, musks, Cistus-labdanum, patchouli, vetiver, vetiveryl acetate and methyl ionone.

Starting with just Oakmoss and Ambergris 162B (substitution Cistus, abs)
Proportions to be smelled on sample strips are:

Oakmoss	9	8	7	6	5
Cistus	1	2	3	4	5

Choosing one of these accords, it is necessary to add a musk note:
6 Oakmoss
4 Cistus
1 Musk Ketone

Smelling a sample strip gives an unpleasant odor so addition of a modifier of intermediate volatility is necessary, a floral Rose note is selected and a touch of an animalic note such as Civet:

Bases
6 Oakmoss
4 Cistus
1 Musk Ketone
Modifier
3 Rose
1 Civet

To this composition is added the Top note which produces the immediate olfactory effect. Considerable flexibility and freedom can be used for selection of the Top note. A selection of bergamot and sweet orange is used in this example. The percentage of each of the three note classifications plays a considerable role in the tenacity of a perfume and the base should represent a large share of the fragrance. This simple accord with the addition of the top notes is show below. Comparisons can also be made for additional ratios of the base notes with a range of 3-4 top notes.

Top Notes (25%)
4 Sweet Orange
1 Bergamot
Modifier (20%)
3 Rose
1 Civet
Bases (55%)
6 Oakmoss
4 Cistus
1 Musk Ketone

The above formulation can be extended and modified by addition of another base note such as Vetiver in a range of relative proportions as done with the starting bases:

	A	B	C	D
Oakmoss	4	6	3	3
Cistus	4	3	6	3
Vetiver	4	3	3	6
Musk Ketone	1	1	1	1

The main characteristic of these four Accords will be the dominant base as Oakmoss in B, Cistus in C and Vetiver in D. Patchouli and Methyl Ionone can also be included, substituted and/or utilized for a dominant characteristic in the accord. The approach offers many possibilities for creating new accords and perfumes as long as the materials selected are from the series of base notes for the accord.

A composition for a simple Chypre perfume developed by further addition of Top notes, Modifiers and Bases is shown below:
Top Notes
Bergamot, Lemon, Linalyl acetate
Modifiers
Jasmin, Geranium, Neroli, Aldehydes (C9-C11)
Bases
Oak moss, Styrallyl acetate, Vetiver, Cistus, Musk Ketone

The modifications are practically endless but the perfume will still be classified in the **Chypre Accord** and Carles' method demonstrates the successive steps for composing a perfume in a particular accord. Carles performed over a thousand experiments with oakmoss alone!

Floral Accords are also commonly utilized in perfume composition and the process is basically the same as set forth above for Chypre by Carles (38,39). However, selection of base notes is more challenging for floral accords. In this case it is recommended to make simultaneous use of base notes, modifiers and top notes, directly creating the perfume. An example of choices to start with for smelling of sample strips for a Jasmine accord is:
Top -   3, 6, 3 - Benzyl acetate (fresh, floral jasmine)
Modifier - 3, 3, 6 - Ylang-Ylang
Base -   6, 3, 3 - Amyl cinnamic aldehyde (fruity, herbal jasmine)

Again many variations of materials can be utilized including Santalol (warm, sweet, woody), Linalool (citrus, floral, sweet), Benzyl salicylate (balsamic, orchid herbal); etc.

Similar approaches can be utilized for creating additional accords based on the perfume pyramid. A sample classic **Fougère Accord** pyramid is composed of (36):
Top - 30% Lavender & Aromatic Notes
Heart - 20% Rose or Geranium Notes
Base - 30% Patchouli or Moss Woody Notes, 10% Coumarin, 10% Salicylates

A sample classic **Oriental or Amber Accord** pyramid is composed of (36):
Top - 30% Citrus & Aromatic Notes
Heart - 20% Floral with Rose & Jasmin Notes
Base - 50% Patchouli, Frankincense, Myrrh, Benzoin, Opopanax, Styrax, Vanilla, Coumarin

Curtis and Williams (43) described additional approaches for development of accords and fragrances in their book "An Introduction to Perfumery." They provided a more detailed approach for creation of fifteen Floral Bases of Muget, Acacia, Carnation, Gardenia, Honeysuckle, Hyacinth, Jasmine, Lilac, Lily-of-the-Valley, Narcissus, Orange Blossom, Rose, Sweet Pea, Tuberose and Violet.

Shown in Table 3.3 for Muget is a list of the of the ingredients which are added as droplets (0.01 g) in a prescribed sequence for creation of the lily-of-the-valley floral base. Sample strips are smelled after each addition, noting the odor effect of the added material. Recommended upper limits for each dosage for the Muget base were provided by the authors and just the final calculated weight percentage of each component is shown in Table 3.3.

Table 3.4 shows the chemicals selected for a few more bases. The Carnation base is started with a high concentration of Eugenol followed by sequential addition of the chemicals shown in the Table. Likewise, the sequential addition of the appropriate chemicals for a Rose base starting with Geraniol is also shown in Table 3.4.

Calkin and Jellinek (15) have also listed Compounding Notes for eight floral accords of Rose, Jasmine, Muget, Lilac, Carnation, Hyacinth, Violet, and Neroli Orange Blossom, with a Basic Formula of Notes for each and additional Selected Variants and Modifiers.

In addition to the odor of essential oils, accords and individual synthetic molecules, there are many simple combinations of materials that can provide particular notes, some of which might not be obvious, especially to the novice perfumer. Carles (38,39) referred to these as "tricks of the trade" and a few of his and some additional combinations are shown in Table 3.5. There are many more and since they form a new or familiar odor they are essentially simple bases with perhaps a few auxiliary notes (19,47,59).

Table 3.3 Creation of a Muget Base

Ingredient*	Wt % of Formula	Odor Effect
Lyral	38.03	Muguet, rather chemical
Linalyl cinnamate	9.13	Smoother, more floral, deeper
Phenyl propyl alcohol	3.80	More rounded, but still chemical
Citronellol	13.69	Better, more like Muget flowers
Alpha Ionone	4.56	More floral
Linalyl acetate	3.80	Fresher
Heliotropine 20%	19.01	Smoother powdery: too much repeat with less
Bergamot oil	1.14	Slightly fresher
Jamine base T109	1.90	Harsher – try better Jasmine Repeat
Lilial	1.14	Improved – better than before adding T109
Citronellyl acetate	1.90	Not much difference
Benzyl acetate	1.90	Gives lift
Jamine base T109	1.90	Harsher – try better Jasmine Repeat
Lilial	1.14	Improved – better than before adding T109
Citronellyl acetate	1.90	Not much difference
Benzyl acetate	1.90	Gives lift

*Droplets (0.01g) added sequentially down the list for creation of the base. Just the final calculated weight percent of each is shown; Reference: Curtis and Williams (43)

Table 3.4 Creation of a Carnation and Rose Bases

Carnation Base Ingredient	Note*	Rose Base Ingredient	Note*
Eugenol	M	Geraniol	T
iso-Eugenyl acetate	M	Geranyl acetate	M
Methyl isoeugenol	M	Phenlylethyl alcohol	T
Benzyl isoeugenol	M	Citronellol	T
Lyral (IFF)	B	Eugenol	M
α-Ionone	M	Phenylacetic acid, 10%	B
Vanillin, 10%	B	α-Ionone	M
Phenylacetaldehyde, 10%	B	Phenylacetaldehyde, 10%	B
Helitropine, 20%	M	Aldehyde C11, 10%	T
iso-Amyl salicylate	T	Aldehyde C8, 10%	T
Benzyl salicylate	T	Citral, 10%	M
Phenylethyl alcohol	T	Aldehyde C12, 10%	T
Geraniol	T	Nerol	T
Citronellol	T	Aldehyde C9, 10%	T
Benzyl acetate	T	Linalool	T
Extension of formula:		Extension of formula:	
Anisaldehyde, 10%	M	iso-Butyl phenylacetate, 10%	T
Phenylpropyl alcohol, 10%	M	Phenylpropyl alcohol	M
Methyl eugenol	T	Geranyl butyrate	T
γ-Methyl ionone	T	Phenylethylisobutyrate, 10%	T
α-Terpineol	T	Trichloromethylphenyl carbinyl acetate, 20%	B
Enhancer examples of Natural Origin:		Enhancer examples of Natural Origin:	
Top – Bergamot oil, Carrot seed oil		Top – Bergamot oil, Lemon oil, Mimosa abs.	
Middle – Clary sage oil, Jasmine abs. Clove bud oil, Orange flower abs.		Middle – Geranium oil, Palmarosa oil, Rose abs., Ylang-Ylang oil, extra	
Base – Benzoin resinoid, Black pepper oil, Peru Balsam oil, Pimento berry oil		Base – Ambergris tincture, Benzoin resinoid, Labdanum resinoid, Vetiver oil	

*T=Top note, M=Middle note, B=Base note; Chemicals are added sequentially down the list (refer to Table 3.3); Reference: Curtis and Williams (43)

Table 3.5 Desired Odors from Combinations of Materials

Odor	Material Combination
Flowers	
Tuberose	Aldehyde C18, Argeol (Santalol) & Celery
Lilac	Styrax (abs) w/ either Hydroxycitronellol or Phenyethylalcohol option - Phenyethylalcohol, Helional, Indole, Clove buds (essence)
Rich Lilac	Styrax (abs), Phenyethylalcohol & Indole
Jonquil	Neroli & Styrax (abs)
Carnation	Benzyl salicylate & Eugenol
Freesia	β-Ionone & Linalool
Orange Blossom	Methyl anthranilate & Linalool
Others	
Bitter Orange	Sweet orange oil & indole
Blood Orange	Sweet orange oil & Ethyl maltol
Fruity	Phenoxyethyl isobutyrate & Dimethyl benzyl carbinyl acetate
Clary Sage	Violet leaves (abs) & Petitgrain
Ambergris	Labdanum, Olibanum & Vanilla
Cucumber	Lavender (abs) & Violet leaf (abs)
Chocolate	Isobutyl phenylacetate & Ethyl vanillin
Coca Cola	Vanillin & natural Cinnamon, Orange & Lime
Metallic effects	Add Ally Amyl Glycolate or Amyl Salicylate

References: Carles (38, 39), C. Burr (47), Ellena (58, 59)

Jean-Claude Ellena (58,59) similarly demonstrated how smells can be reduced to a few constituents by subjecting your nose to a minimum of two test blotters with different notes at 5% concentration in ethanol. The blotters are dipped and smelled separately by wafting, then put together and the combination aroma again smelled by wafting. Table 3.6 shows some of the new odors obtained from the combination of blotters smelled together.

Table 3.6 Desired Odors Obtained by Wafting Combinations of Materials

Odor	Material Combination*
Amber	Vanillin & Labdanum (absolute)
<u>Sweets</u>	
Cotton Candy	Vanillin & Ethyl Maltol
Carmel	Tonka bean (absolute), Vanillin & Methyl cyclopentenolone
<u>Flowers</u>	
Gardenia	C-18 Aldehyde Prunolide, Styrally acetate & Methyl anthranilate
Hyacinth	Phenyl ethylalcohol, Benzyl acetate & Galbanum
Jasmine	Benzyl acetate, Hedione, Clove bud oil, Indole & Methyl anthranilate
Lily	Benzyl salicylate, Phenyl ethylalcohol, Methyl anthranilate
<u>Fruit</u>	
Apples	
Green	Fructone, Benzyl acetate & cis-3-Henenol
Yellow	Fructone, Hexyl acetate & Benzyl acetate
Red	Fructone, Allyl caproate & Hexyl acetate
Cherry	β-Ionone, Heliotropine & Benzaldehyde
Fig	Stemone & γ-Octalactone
Grapefruit	Sweet orange & Rhubofix
Mango	Ionone, C-14 Aldehyde & Black currant (absolute)
Pear	Fructone, Hexyl acetate & Geraniol
Peach	Fructone, C-14 Aldehyde & Black currant (absolute)
Pineapple	Allyl hexanoate & Ethyl maltol
Raspberry	Fructone, β-Ionone Geraniol & Ethyl maltol
Strawberry	Fructone & Ethyl Maltol

*Samples are smelled by wafting together sample strips dipped in the different ingredients. Reference: Ellena (58, 59)

Composing Perfumes

Most perfumes are begun with a Brief, which is a blueprint, the visions and underlying story of the fragrance. It can be envisioned by the perfumer, the client or the perfume house. Briefs can be simple or very detailed and are often a conceptual description of an imagined scent such as a "Walk in a Field of Flowers" or "Fresh Splash of Water from the Mediterranean" or "Smell of Autumn."

Inspirations for fragrances can also be of a special moment, song, flavor or color, with consideration of texture as smooth, silky, metallic or earthy. Jean-Claude Ellena used simple themes for some of his excellent perfumes such a "tea" for *Eau Parfumée au thé Vert (Bulgari)*, "ginger & water" for *Un Jardin Après la Mousson (Hèrmes)*, "flour odor" for *Bois Farine (L'Artisan Parfumer)*, "fig leaf" for *Un Jardin en Méditerranée (Hèrmes)* and "green mango" for *Un Jardin sur le Nil (Hèrmes)* (58). Some companies develop very detailed briefs which include marketing campaigns, customer profiling, raw material costs, design concepts, development schedule (up to 18 months) and packaging.

There are a number of classification schemes and approaches for creation of new fragrances. The perfumer Jean-Claude Ellena classified odors into nine categories, each with subgroups, with examples given, as follows (58,59):
Flowers (Rose, White, Yellow, Exotic or Spiced, Anise)
Fruits (Citrus; Orchard Fruits, i.e fructone; Soft Fruits, i.e black currant)
Woods (Sandal, Patchouli, Vetiver, Cedar, & Lichen, i.e. oak moss)
Grasses (Green, i.e. galbanum; Aromatic, i.e. lavender; Aniseed, i.e. basil)
Spices (Cool, i.e. cardamom; Hot, i.e. cinnamon)
Sweet Products (Vanillas, Coumarins, Musks)
Animal Products (Ambers, i.e. labdanum; Castoreums; Civet, i.e. indole)
Marine (calone)
Minerals (aldehydes)

Ellena (58,59) further described that a perfumer makes up a collection of fragrances according to the following categories: Conceptual ideas, Technical Quality, Cost, and Reliability & Availability.
The **Conceptual** ideas include:
Emotional (aggressive, erotic or hedonic sensations),
Sensorial as taste (sweet, salty, bitter, ec.), touch (harsh, soft, etc.), physical (volume, mass) and dimension (flat, thick)
Descriptive (floral, spicy, grassy, marine, etc.).

The **Technical Qualities** are:
Tenacity or substantivity,
Intensity (levels of perception by diluting in stages until the end of olfactive sensation),
Volume (placing a blotter with a drop of diluted odorant in a jar for one hour),
Cost, and the **Reliability & Availability** in terms of geographical and economic positions.

Perfume Blending

The Perfumer develops a new fragrance by blending sometimes hundreds of natural essential oils and synthetic chemicals, although it is preferable to have fewer components in the final fragrance. A skilled "nose" can identify thousands of different scents and a typical perfumer will have over 1,000 choices in his/her fragrance work station. Perfumes can contain hundreds of notes but most perfumers now tend to limit the ingredients to less than one hundred and even less than fifty. This is because the slightest change in the relative proportions of perfume components can result in a significantly different product (58,59,63-66).

Natural essential oils generally provide sweet, floral and rather heavy odors while the synthetics add strength and sparkle to blend, especially the aldehydes which provide for a more sheer, clean and lighter perfume. The choice of natural vs synthetic components is dependent upon both quality and cost but, as previously pointed out, synthetics comprise the majority of the composition of contemporary perfumes.

A few materials are found in almost every composition, notes of rose, jasmine, violet, wood, patchouli, sandalwood, vanilla and synthetic musk. Some other constituents found in almost all perfumes are ambroxan, phenylethyl alcohol, citronellol, coumarin, hedione, heliotropin, hydroxycitronellal, iso E super, ionone, lilial, methyl ionone and salicylates, all materials with linearity-specific character (59). Additional common notes that are part of accords include vetiver, tree moss (formerly), violet leaf, isoquinolines, aldehydes and birch tar. The chemicals are further described in the chapter on Chemistry of Odorants.

It usually takes many months to years once the brief has been established for selection of the final fragrance (47,58,59,63-66). With the brief in hand the perfumers will conceive of a possible composition for the perfume based on their extensive knowledge of the natural and synthetic oils in their personal repertoire, including what is known of interactions of the materials in a fragrant mix. This is where the extensive training about the notes, bases and accords is utilized by the perfumers for formulating a conceptual fragrance.

Roudnitska (63-66) expounds on the idea that intuition for creation of new fragrances must be realized with the training and extensive knowledge of the natural and synthetic oils, but what he describes is our modern understanding of creativity and the factors that favor a creative mind. These factors include attributes such as personality, intelligence, style of thinking, motivation and environment. Creative people are very curious, open-minded, risk takers, high energy, persistent, imaginative and subject to daydreaming (67,68). With his extensive knowledge of perfume ingredients, Jean Carles was able to formulate his exceptional fragrance, *Ma Griff*, even after he had lost his sense of smell (anosmia)!

The perfumer will often make a list of different odors that would express the concept as top, middle (modifier) and base notes and would note relative proportions to use in an initial first trial. However, many modern fragrances are composed using fragrance bases, that is, previously created modular bases of specialized scents with multiple ingredients as previously described. The modular bases are usually concepts such as "fresh cut grass," "sour apple," "floral" or even more specific such as for flowers that are not extracted such as gardenia or hyacinth. Modifiers such as fruit esters can be included in a floral base for a fruity floral or calone and citrus for a fresher floral (15,16,37,58,59,).

Blender chemicals are added to smooth out the transitions between the different layers or bases, with linalool and hydroxycitronellal utilized in many fragrances for this effect. Fixatives such as resins, wood scents and amber bases are added to bolster and support the primary scent (5,16,37).

Accessory ingredients can also be incorporated. These are materials that cannot usually be used in large amounts since even trace presence in a formulation can impart a unique cachet. Examples are aldehydes (C12 MNA and C13), styrallyl acetate, isobutylquinoline, galbanum, Ambrocenide (fixative for citrus notes), Cosmone (musk, volume lifter), Ultrazur (lift, volume & diffusitivity), Okoumal (adds twinkle to woods) and ambergris (expensive!) (38,39,49). However, addition of ingredients in small amounts is not to be taken as a strict rule in perfumery, especially for contemporary fragrances. Modern fragrances provide contrasts with sharp olfactory values and there are no incompatibilities in perfumery (see section on Horizontal Perfumes below) (38,39).

Ellena's approach for a fragrance of sweet pea flowers was to sketch out a simple formula of just seven ingredients (58,59). He perceived the aroma of sweet pea to be between that of roses and orange blossoms. After five trials, changing proportions and with the addition of a note of carnation, he had the starting formula for the perfume shown below.

Sweet Pea (5% in Ethanol)
Phenylethyl alcohol   200
Paradisone   180
Hydroxycitronellal   50
Rhodinol   30
Acetyl eugenol   15
Orange blossom   15
Cis-3-Hexenol   5
Phenyl acetic aldehyde 5 (at 50% concentration)

As an initial approach the perfumer would gather the bottles of all the notes and bases initially conceived for the new perfume and start by applying droplets of the different odorants onto blotters and sample strips to test interactions. However, materials of low vapor pressure could appear weak on the blotter in undiluted form but may exhibit unexpected effects in compositions on the skin. Mixing the materials in specific relative proportions with ethanol is an important step.

Contemporary perfumers use a computer program for selecting materials from a list of those available, both within their own company and on the market, to compose a formula. The programs are designed to send alerts for ingredients limited by regulatory concerns and provide replacements for expensive materials. Additional applications of computer and artificial intelligence to perfumery are described in a later section.

The perfume notes and bases are blended and matured for several days to allow the ingredients to harmonize. The perfume concentrate is then macerated in alcohol for weeks to months to stabilize the fragrance. This is followed by filtering to remove sediment and particles before transferring to the perfume bottles (58). Many perfumes also include colorants to improve the marketability, glycol diluents and antioxidants for improvement of perfume shelf life (14-16, 41-44).

Although smelling strips are utilized for evaluation of notes, simple bases and accords, Jean Carles was adamant that perfumes should be evaluated in vapor form by spraying with a simple device. He suggested vaporizing the perfume for 5-7 seconds in a room, closing the room and re-entering the room after 2-3 minutes for the olfactory evaluation. The vaporization produces the true, fully developed scent and allows for immediate modification of the fragrance formulation. This approach allows for a large number of evaluations compared to the use of smelling strips. However, it should be noted that smelling strips are indispensable for evaluation of perfumes.

Creation of Fragrance *Un Jardin Sur Le Nile*

Chandler Burr (47) provided further insight into Ellena's approach to perfume creation in his book "The Perfect Scent." He followed Ellena's progress for the development of the fragrance *Un Jardin Sur Le Nile* for Hermès from the brief to the final perfume selection. Ellena was Hermès in-house perfumer and had recently composed the "fig leaf" aroma fragrance *Un Jardin en Méditerranée* for the company following their "River" theme.

The "River Nile" was the entire brief for the new fragrance as well as specifying the name of the perfume *Un Jardin Sur Le Nile*. Several Hermès representatives and Ellena travelled to Egypt to garner impressions for the perfume. They would visit the Kitchener Garden in Aswan, the markets in the city and a Nubian village smelling the air, blossoms and fruit along their route.

Before the trip Ellena had already conceived of a perfume inspired by Egypt with heavy aromas of incense, strong jasmine and wood smoke, but this was dramatically altered after the experience in Aswan. Some of the aromas they encountered included nasturtium (green/watercress scent), lantana (banana/passion fruit) and acacia (soft frangipani scent) in the garden; lotus root (peony/hyacinth) and jasmin sambac (indolic/animalic) in the market. But it was on their stroll to the Nubian village that Ellena was inspired by the scent of green mangoes, an ephemeral fresh and rich aroma. It also exhibited a hint of apricot and grapefruit typically associated with the synthetic, acetone. In contrast, ripe mango has a tropical fruity, peach/plum type odor (47).

Ellena wrote down a basic formula for the perfume comprised of thirteen ingredients. His approach was to simulate green mango, that is, to create a fantasy of green mango. Three vials of fragrances were prepared, one inspired by the Green Mangoes, another with more Lotus (root) and a third more Woody with a lot of incense. All three contained the scent of carrot, which was detected from the green mangos in Aswan. The trial perfume with more Lotus was chosen. Lotus root has the subtle scent of lotus, sandalwood, cedar & musk while Lotus flower is floral with watery, aqueous qualities, light and ethereal with a slight sweet tonality. Some changes were requested to make the Lotus formulation a bit less harsh, by reducing the grapefruit odor and increasing the presence of green mango.

Some of the ingredients in the samples included (47):
Bitter orange (simulate freshness of green mangos)
Synthetic grapefruit (for mango's acidity)
Lotus root (subtle scent of lotus, sandalwood, cedar & musk)
Carrot (green, sweet, herbaceous, minty nuance)
Rosin (earthy spice, sweet, gingery with musky undertones)
Hay (possibly coumarin)
Acetone (adds jolt, smells of nail polish remover at high concentration)
Sycamore (nutty, woody, earthy)
Citrus (grapefruit)
Calamus (warm, spicy-cinnamon, woody with sweet undertones)
Methyl ionone (iris odor)
Audépine (Hawthorn, sweet with animalic nuances, rubbery/clove scent)

Opopanax (resinous odor) and another synthetic mango odor in the original sample were eliminated because the opoponax created a mushroom odor in the juice and the mango synthetic was perceived as apricot. These adjustments illustrate how materials in a composition can interact to produce aberrant odors.
To enhance the scent of mango Ellena added:
Hedione (simulates jasmine), Methyl anthranilate (sweet fruity, concord grape smell, musky/berry nuance) and Neroli oil (from orange blossoms)

Ellena then created three more variants of the original, but they all were too citrusy. He lowered the grapefruit and added trans-2-Hexenal (green, fruity/apple) and in further modifications he added a more Flowery/Fruity note, Honey absolute and Pine Incense. He also further lowered the citrus and bitter orange. The fragrance now had thirty-two ingredients and smelled immediately of green, unripe mango, Ellena's goal, and with a fresh, green, sweet, fruit sillage. The stages of fragrance development described for Ellena's perfume demonstrate the many perturbations a perfume goes through and the time it takes for final acceptance and marketing.

The final perfume has the lightness and luminosity of modern perfumes with a fragrance of citrus, green, floral, tropical, fruity and sweet. A published perfume pyramid for this fragrance is (11):
Top: Grapefruit, Green Mango, Tomato, Carrot
Heart: Bulrush, Lotus, Orange, Hyacinth, Peony
Bottom: Musk, Iris, Incense, Labdanum, Cinnamon

Chandler Burr (47) also revealed the formulation for "hugely successful commercial luxury perfume" the name of which he was obligated not to disclose. He was given the fragrance to assess and had it analyzed by gas chromatography. It is useful to see the range of thirty-three natural and synthetic compounds and bases in the juice as shown Table 3.7 (listed from highest to lowest concentration). It was surmised to be the almost complete formulation for *Un Jardin Sur Le Nile* in an internet publication (reddit.com), although the formulation differs to an extent from the progression to the final fragrance as described above.

Table 3.7 Composition of Unknown Luxury Fragrance

Constituent	Content, %	Odor
Ethylene brassylate	30.4	Sweet, musky, powdery
Dipropylene glycol	18.5	Carrier oil, holds notes in harmony
Cassis base 345B	9.1	Green, floral, lychee fruity
Hedione high cis	8.9	Jasmine, magnolia, sweet green
Nerolidol	5.6	Floral green, waxy, woody
Sandiff	5.3	Sandalwood, clean, sweet, woody
Magnolan	3.3	Floral, green, reminiscent of magnolia, peony, grapefruit
Bergamot, Italie	3.0	Citrus, top note
Phenyl ethyl alcohol	2.0	Floral rose, rose water
Linalool	1.3	Citrus, sweet lavender
cis-3-Hexenol, 10%	1.3	Leaf alcohol, intense grassy green
Linalyl acetate	1.2	Sweet, green pear, floral, citrusy
Ambrettolide	1.1	Musky, powerful, warm (Givaudan)
β-Ionone	1.0	Violet
Helional	0.9	Floral, muget, melon, watery
Lemon, Italian	0.9	Citrus, lemon
Ethyl acetoacetate	0.9	Fresh, fruity, green apple
Geraniol	0.5	Sweet, floral rose
β-damascone, 10%	0.5	Rose, plum, black currant
Grapefruit	0.5	Citrus, grapefruit
Methyl pamplemousse	0.5	Fresh citrus, grapefruit (Givaudan)
Ally amyl glycolate, 10%	0.4	Fruity, pineapple, spicy green
cis-3-Hexenyl acetate	0.4	Sharp fruity, green banana
Geranium, bourbon	0.4	Rosy green, spicy
Geranyl acetate	0.4	Floral, fruity, rose
Cardamom, Guatemala	0.3	Warm, spicy, camphoraceous
Dimethyl benzyl carbinyl butyrate	0.3	Fruity, plum, herbaceous
Phenoxyethyl isobutyrate	0.3	Fruity, rose-honey
Fleur d'oranger, F175SAB	0.2	Orange flower blossoms
Mousse synthétique, 10%	0.2	
Tuberose base	0.2	Floral, creamy, fruity-peachy
Carvone, Laevo, 10%	0.1	Sweet minty
Reseda body, 10%	0.1	Floral, hyacinth, narcissus

Reference: Burr (47), Surmised to be close to the formulation for the fragrance, *Un Jardin sur le Nil*, by BostonPhotoTourist, Reddit.com

The final formulation attests to the continued effort by the perfumer to find the optimum combinations of ingredients to achieve the superb resulting fragrance. The perfume has a strong but smooth ripe, fruity aroma with excellent diffusion and persistence. The chemical constituents in three other classic and popular fragrances are shown in Table 3.8 and many of the chemicals in the formulations are discussed in the next chapter on the Chemistry of Odorants (69). A list of commonly utilized chemicals in perfumery and their odors is provided in Appendix I.

Table 3.8 Some Chemical Constituents in Three Popular Fragrances

Chemical	Coco (Chanel)	Happy (Clinique)	Curious (Britney Spears)
Avobenzone	X		
Benzyl salicylate	X		X
Benzyl benzoate			X
Benzyl acetate			X
Butyl acetate			X
T-Butyl alcohol		X	X
2-tert-Butyl cyclohexanol			X
Cashmeran	X		
Citral	X	X	X
Citronellol	X	X	X
Coumarin	X		
Diethyl phthalate			X
Ethylene brassylate	X	X	
Eugenol			X
Farnesol			X
Galaxoide	X	X	X
Geraniol	X	X	X
Hedione	X	X	X
Hexylcinnamal	X		
Hexylacetate			X
Hydroxcitronellal	X	X	X
Isoeugenol			X
Isodecane			X
α-Isomethylionone		X	
Isoamyl butyrate			X
Limonene	X	X	X
Linalyl acetate	X		
Lilial	X	X	
Linalool	X	X	X
Linalyl anthranilate	X		
Lyral	X	X	
Octinoxate	X		
Methyl cinnamal			X
Myrcene	X	X	X
Octisalate	X		
Phenylethyl alcohol	X	X	
γ-Pinene	X		
Terpineol			X
γ-Terpinene	X		X

Table 3.8 (Continued)

Chemical	Coco (Chanel)	Happy (Clinique)	Curious (Britney Spears)
Tonalide		X	
Trans-β-Ionone	X		
cis-2,6-Dimethyl-7-octenol	X	X	X
cis-2,6-Dimethyl-2-6-octadiene	X		X
3,7-Dimedthyl-1,3,7-Octatriene	X		
Diethyl phthalate	X		
Parcymene	X		X
Tromethamine		X	
Isopropyle myristate		X	

Reference: Disclosed Chemicals: Perfumes (69)

Perfume Overdosing

Overdosing or inclusion of a high concentration of a particular ingredient in a perfume is commonly utilized in fragrance formulations. The overdosed constituent provides the perfume with special impact, character and identity. The famous *Chanel No.5* contains an especially high concentration of jasmine which is even detected initially as top note, at the heart and in the dry down as a base note. *L'Origan (Coty)* has a substantial overdose of methyl ionone and dianthine (20%, Carnation base). *Chypre* by Coty is overdosed with both oakmoss and civet, providing the classic character of the chypre accord, along with a balance of floral notes.

Some other examples of overdosed fragrances are *Fougere Royale (Hougibant)* with coumarin, *Vetiver (Guerlain)* with about 20% vetiver, *Milliseme (Creed)* with damascones, *Trèfle Incarnat (L. T. Piver)* with amyl salicylate (steely odor), *Après l'Ondée (Guerlain)* with anisic aldehyde (mimosa/frangipani) and *Aventus (Creed)* with Ambrox. The use of overdosing is also notably employed for creation of "Horizontal perfumes." (13-16,59)

Horizontal Perfumes

Another approach to creation of perfumes was developed by Sophia Grojsman who did not follow the olfactory pyramid. Grojsman created what are termed "Horizontal Perfumes" or "Linear Perfumes" using her signature Grojsman Accord with no clearly defined top, middle and base notes. Fragrances with this accord impart essentially the same aroma from start to finish many hours later. Realizing the immediate full impact of the perfume is a great advantage for retail counter sales (13-16,19).

In her pioneering approach, Grojsman was the first to incorporate Iso E Super (Isocyclemone E) in her formulation of *Tresor* for Lancôme in 1990. The Iso E Super created exceptional sillage in the perfume and was used in concert with a several other synthetic materials to produce the dramatic effect of the fragrance. Her accord consisted of 21% Galaxolide (warm musk), 18% Iso E Super (cedar), 18% Methyl ionone (violet), and 6% Hedione (lemon) and comprised 80% of *Tresor*. This warm Floral-Oriental perfume has an aroma of rose, muget and lilac with the sparkle of peach and apricot and continues to be popular. Many contemporary perfumes contain Iso E Super and it has had a profound effect on the path of fragrance development (13-16,19,59).

Grojsman, currently Vice President for IFF, is one of the most prolific perfumers in recent times. She has created over 50 perfumes for numerous perfume houses since the 1980s including *Tresor* and *Lalique* for Lancôme, *Paris* and *Yvresse* for Yves Saint Laurent, *Vanderbilt* for Gloria Vanderbilt, *White Linen* and *Beautiful* for Estée Lauder, *Diamonds and Rubies* for Elizabeth Taylor, *Eternity* for Calvin Klein and *Bvlgari Pour Femme* for Bvlgari, to name a few (17,19).

Computer Technology & Artificial Intelligence (AI) in Perfumery

Many perfumers use proprietary computer spreadsheets for collection of fragrance data, however there are a number of commercial databases available for purchase. Leffingwell & Associates offer a "Perfumes & Fragrances Database (PFC 2002)" with classification of 5,500 original perfumes and other cosmetic products with specifications for type, odor, brand name, year and creator, if known. It includes a semi-qualitative indication of the composition of over 4,000 modern perfumes with the possibility for searches of name, type, odor and composition. The type of information available in a typical database is shown in the Figures 3.1 & 3.2.

As with intellectual processes of humans, artificial intelligence is the use of digital computers and robotics to perform intelligent tasks. Computers can be programmed to carry out very complex tasks with great proficiency for applications as diverse as medical diagnosis, voice recognition and the pervasive computer search engines. The approach has now been applied to formulations of perfumes. However, the AI programs are not without human interaction and are utilized in conjunction with skilled perfumers for optimization of fragrance formulations (70-75).

Kirchoff et al. (73) described the use of automated process research (APR) technology for optimization of the relative proportions of components in fragrance formulations. Libraries of perfume formulation can be generated to facilitate rapid testing of new components and the combination of the ingredients can then be optimized via statistically designed experiments.

Givaudan has developed an AI program named "Carto" which eliminates the need for spreadsheets and analyzes multiple data sets from their library of market research, consumer data and historical fragrance formulas to arrive at a fragrance recommendation. The company's Odor Value Map, "Eve," is used to create products based on the requests of the customer. The program works out the mathematical proportions and a linked robotic compounding machine produces the sample fragrance in seconds. The approach has essentially reduced hundreds of trials down to just a few (75).

The German perfume house, Symrise, partnered with IBM to develop another AI program "Philyra" utilizing their huge database of hundreds of thousands of fragrance formulas, fragrance families, thousands of raw materials, previous combinations of raw materials and historical success data to generate new combinations of fragrance formulations that fit specific design objectives.
For example, the system includes algorithms that learn and predict alternative raw material complements and substitutes, human response such as pleasantness and gender appropriateness, and novelty of the fragrance by comparing to commercially

available fragrances.

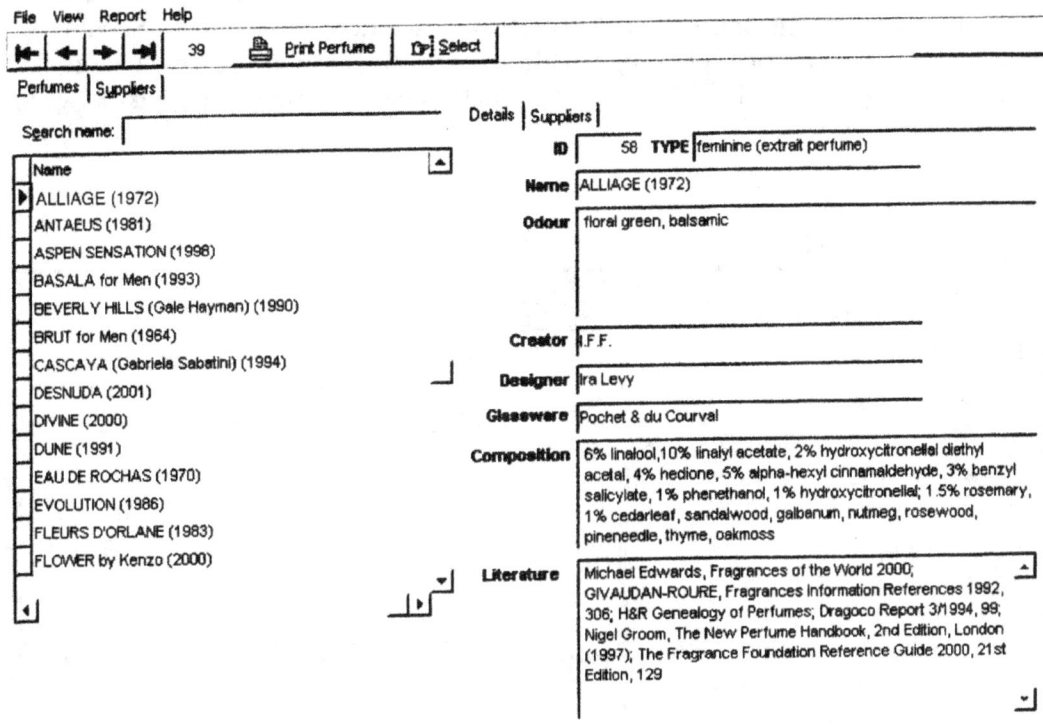

Figure 3.1 Sample of perfume & fragrance classification database for the fragrance, *Alliage* (34).

Symrise has used Philyra to create fragrances for the Brazilian company, O Boticario, tailored to their specific consumer demographics and personality. The AI program produced a fougère fragrance, a herbaceous, mossy type perfume, which the company bottled and successfully sold as *Egeo on You* launched in 2019 and a similar feminine fragrance, *Egeo on Me* (70). Symrise also plans to introduce their AI program into their perfumery school to help train the next generation of perfumers.

More recently Symrise used Philyra to create the *Scent of Berlin*, the scent of a city. They ran an interactive and collaborative project with the citizens of Berlin, Germany, by having participants from the city rate five different fragrances based on their preferences. With the assistance of a perfumer, these "core fragrances" produced by Philyra, were modified each week based on the participants emotional responses to the scents and from answers to questionnaires. The final computer scent was unveiled in June, 2019 as *#Berlin 3.0*.

The French perfume house, Ex Nihilo, similarly utilized their AI robotic program, "Osmologue," to create specially customized fragrances, by altering their existing fragrances based on customer's preferences. Firmenich has taken this type of approach even further through the use of their new nanotechnology, labelled

"ScentMove" (74). MRI scans of brain activity are recorded from individuals smelling different perfumes providing an "Emotion Evaluation Model" based on nine emotional dimensions across 25 commonly used descriptive terms. Scentmove then reveals patterns corresponding to different emotional profiles, which along with cultural background and experiences, enables them to establish a link between emotion and context.

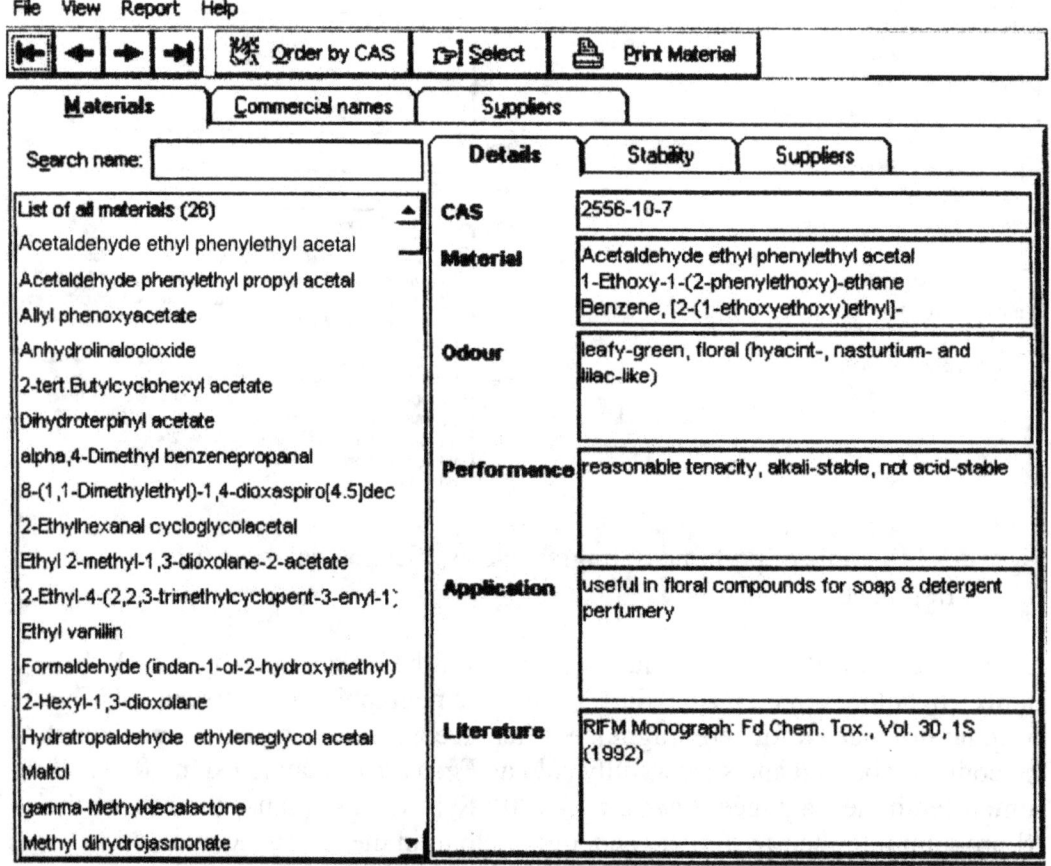

Figure 3.2 Sample of perfume & fragrance classification database for chemical, acetaldehyde ethyl phenylethyl acetal (34).

The company, Algorithmic Perfumery (Scentronix), has developed an AI program that allows the customer to create their own personal perfume. The customer first enters some personal information into the program, such as personality traits and scent preferences, which the program uses to create three sample fragrances in as little as five minutes. It may not produce the ultimate fragrance with this approach but the AI program has proven to be a useful learning approach for the customers (70).

It has been necessary to modify the formulations of many classical and modern perfumes due to health and environmental regulations. Firmenich has also created an AI program labelled "Muse" which can re-engineer a perfume formulation to make it more eco-friendly. The program offers alternative safe chemicals for substitution of the target constituent.

Electronic sensing, "e-sensing" or "e-nose," technology has made significant progress in recent years. It is already in use in the fields of food, agriculture, and the environment. It has the capability of recognizing scents with sensor arrays and pattern recognition. By comparing the volatile compound's fingerprint to a database, the instrument can recognize the new scents. While a graduate student at the Monterrey Institute of Technology, Blanca Lorena Villarreal, developed an electronic "nose" robot which can recognize odors as well as track the source. She is continuing to develop algorithms to widen the variety of odors that the robot can recognize.

AI will not replace the use of perfumers in the near future but will provide valuable assistance for creation of new fragrances and development of customized perfumes. AI also helps to eliminate cultural influences and personal error. Since AI can sense patterns and suggest alternative ingredients, it improves the chance of creating unique and high-quality fragrance products (70-75).

Health and Safety Considerations

An increasing number of ingredients are being restricted or outright banned for use in fragrances. The fragrance industry has had a self-regulating body, the International Fragrance Association (IFRA), in place since 1973. It was set up to represent the collective interests of the industry and promote the safe use of fragrances around the world. IFRA monitors the use of ingredients that produce harmful effects such as skin irritation, allergenic effects, photo effects, neurotoxicity and reproductive effects and advises on restrictions. The restrictions and banning of substances by IFRA are based on evaluations by an independent organization, the Research Institute of Fragrance Materials (RIFM). The IFRA website is regularly updated with new information on restrictions and banned materials (76).

Perfumes can contain additional chemicals such as preservatives which can cause concern as well as some classes of compounds. Parabens are a common preservative used in many fragrances and are suspect since they can interfere with production and release of hormones. Phthalates, also used as a preservative in commercial perfumes, are known carcinogens and health effects include damage to liver and kidneys, birth defects and reproduction problems. Synthetic musks may disrupt hormones and have been found in fat tissue and breast milk (76-80).

The European Union has been more proactive for restrictions and banning of ingredients than in the USA. A recent partial ban has been placed on oakmoss produced from the fungal lichen. Oakmoss is the primary component of chypre perfumes where it provides smoothness with earthy, woody notes and acts as a fixative. It is an important component of legendary perfumes such as *Chanel No.5*, *Miss Dior* and *Mitsouko*. However the ban was not total for oakmoss, it was for just two chemical components, atranol and chloratranol, in the oakmoss. Unfortunately, the altered oakmoss is a much lighter and less vigorous scent (79).

The list of 26 allergenic substances that have been utilized in perfumery are shown in Table 3.9. These substances must appear on the label of the product when present in the finished formula at certain thresholds. Table 3.10 shows a few of the 130 restricted ingredients by IFRA and their source. Restricted ingredients must be used below a specified concentration in the final product and are present in many classic and modern fragrances (76-80).

Table 3.11 shows some of the 84 Prohibited ingredients and the cause for their prohibition. Prohibited ingredients cannot be used at all with the exception of 0.1 % of "unavoidable impurities" in essential oils (76-80).

Table 3.9 The 26 Allergenic Substances Utilized in Perfumery

INCI Name	Examples of Sources*
Amyl cinnamal	Synthetic
Amycinnamyl alcohol	Synthetic
α-Isomethyl ionone	Synthetic
Anise alcohol	Anise & Vanilla (Tahiti)
Benzyl alcohol	Balsams + Jamine, Black currant
Benzyl benzoate	Balsams + Jamine, Ylang-Ylang
Benzyl cinnamate	Balsams, Copahu
Benzyl salicylate	Propolis
Butylphenylmethylpropional	Synthetic
Cinnamal	Cinnamon, Hyacinth, Patchouli
Cinnamyl alcohol	Hyacinth
Citral	Citrus, Eucalyptus
Citronellol	Lemon grass, Orange
Coumarin	Tonka beans, Sweet clover
Eugenol	Cistus, Ylang-Ylang, Camphor Patchouli, Lemon grass
Farnesol	Rose, Neroli, Ylang-Ylang, Tolu Balsam
Geraniol	Rose, Neroli, Geranium, Ylang-Ylang
Hexyl cinnamal	Synthetic
Hydroxycitronnellal	Synthetic
Hydroxyisohexyl 3-cyclohexene carboxyaldehyde	Synthetic
Isoeugenol	Citronella, Ylang-Ylang
Limonene	Lemon, Orange, Juniper, Neroli Bergamot, Ylang-Ylang
Linalool	Lavender, Pine, Geranium, lemon Ylang-Ylang, Rosewood
Methyl 2-octynoate	Synthetic
Oakmoss (*Evernia prunastri*)	Oak moss extract
Tree moss (*Evernia furfuracea*)	Tree moss extract

*Essential Oils except Balsams. Most of the ingredients can be produced synthetically except for mosses. References: International Fragrance Association (76), Meakins (80)

Table 3.10 - IFRA Restricted Ingredients

Acetylated Vetiver oil	Grapefruit oil expressed
α-Amyl cinnamic alcohol	trans-2-Hexenal
α-Amyl cinnamic aldehyde	Hexahydrocoumarin
Angelica root oil	Hexyl salicylate
Benzyl alcohol	Hydroxycitronellal
Benzyl cinnamate	Jasmine absolute (*grandiflorum*)
Benzyl benzoate	Longifolene
Benzaldehyde	Lemon (cold) & Lime Oils expressed
Benzyl salicylate	Melissa oil (genuine *Melissa officinalis*)
Bergamot oil expressed	Methyl eugenol
Bitter orange peel oil expressed	Methyl heptine carbonate
Carvone	α-Methyl cinnamic aldehyde
Cinnamic alcohol	Methyl ionone, mixed isomers
Cinnamic aldehyde	o-Methoxycinnamaldehyde
Cinnamyl nitrile	Methoxycoumarin
Citral	Opoponax
Citronellol	Peru balsam
Citronellal	Rose ketones
Coumarin	Safrole, Isosafrole and Dihydrosafrole
Cuminaldehyde	Styrax
Dihydrocoumarin	Tea leaf absolute
Eugenol & Isoeugenol	Treemoss extracts
Farnesol	Oakmoss extracts
Farnesal	Ylang ylang extracts
Geraniol	

Reference: IFRA (76)

Table 3.11 IFRA Prohibited Ingredients

Acetyl isovaleryl (S)*
Amylcyclopentenone (S)
Carvone oxide (S)
Cinnamylidene acetone (S)
Costus root oil, absolute and concrete (S)
Cyclamen alcohol (S)
Dihydrocoumarin (S)
Fig leaf absolute (P, S)
Geranyl nitrile (S)
trans-2-Heptenal (S)
trans-2-Hexenal diethyl acetal (S)
Hexahydrocoumarin (S)
6-Methylcoumarin (P)
7-Methylcoumarin (P)
p-Methylhydrocinnamic aldehyde (S)
Musk - ambrette, moskene, tibetene, KS, α, xylene (S, N)
Phenyl benzoate (S)
Quinoline (S)
Savin oil (S)
Verbena oil and absolute (S, P)

*S = Skin Sensitization, P = Phototoxicity, N = Neurotoxicity
Reference: International Fragrance Association (76)

Chapter 4. Chemistry of Odorants

When the famous Nobel Prize Winning Physicist, Richard Feynman, was asked by a journalist, "What single sentence would best encapsulate all science so far if it were to be the sole surviving scrap of all we knew," He replied, "The World is Made of Atoms." If he had been allowed a second sentence, he might have added "And the atoms bond together to make Molecules." The basis of life and perfumery are derived from organic molecules formed from mainly carbon, hydrogen and oxygen.

Basic Chemistry for Odorants

Perfume odorants are either natural (i.e. essential oils) or synthetic volatile organic compounds and can be placed in two general categories of <u>Aliphatic Compounds</u> which occur as Terpenes in plants and <u>Aromatic Compounds</u> which also occur naturally in plants. Both categories of chemicals can be synthesized from petrochemicals.

A brief description of the structure of organic molecules is given to provide a basis for the discussion of the chemistry of both natural and synthetic fragrance chemicals. The carbon atom (C) is the fundamental unit of organic chemistry and when bonded into chemical molecules provides the basis of all essential oils and synthetic substitutes. The carbon chains also have additional atoms bonded to the structure including hydrogen (H) and oxygen (O) and sometimes sulfur (S), nitrogen (N), phosphorus (P) and halogens (chlorine (Cl), bromine (Br), etc.).

Some Simple Rules are necessary for construction of organic molecules.
Carbon can have only four bonds,
Hydrogen one bond, and
Oxygen two bonds.

Aliphatic Hydrocarbons

Aliphatic Compounds are chains of carbon atoms bonded to hydrogen as straight, branched or cyclic. Two simple hydrocarbons are Propane and Hexane:

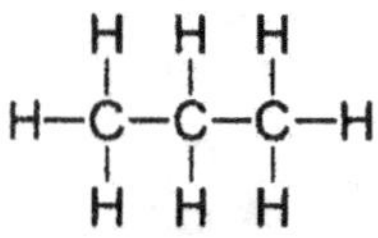

Propane 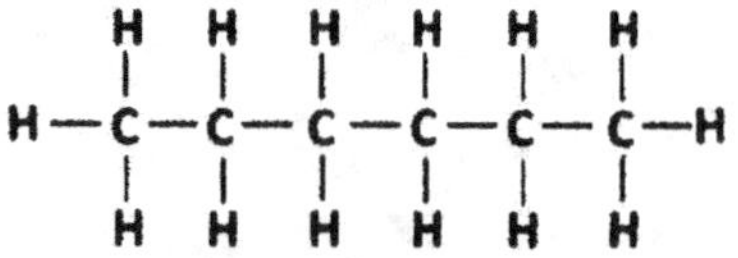   Hexane

Note that all the carbons have four bonds and the hydrogens just one bond. The chains can be extended to quite long structures. The structure can be simplified and represented by a squiggly line where the ends and the angle points represent the carbon atoms bonded with hydrogen as shown below for hexane.

A branched carbon chain structure is shown below for hexane as 2-methyl-Hexane. Note at the branch point a carbon bond has replaced the hydrogen (H), so carbon still has just four bonds.

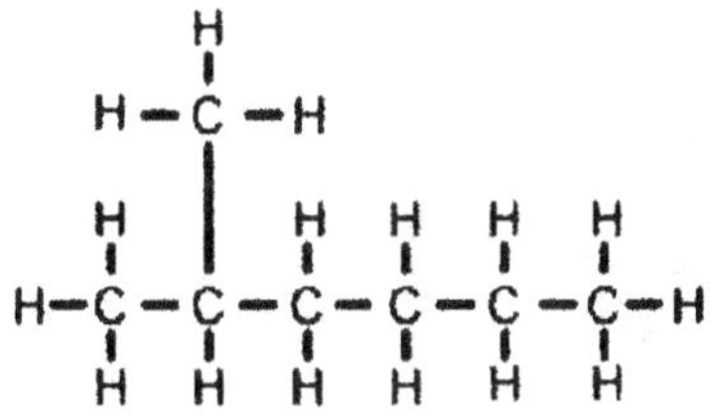

2-methyl-Hexane

The chains can also have carbon with double bonds as shown below for Hexene. Note that the suffice is now **-ene** instead of **-ane**. Also note that hydrogen atoms have been eliminated so that carbon still has only four bonds.

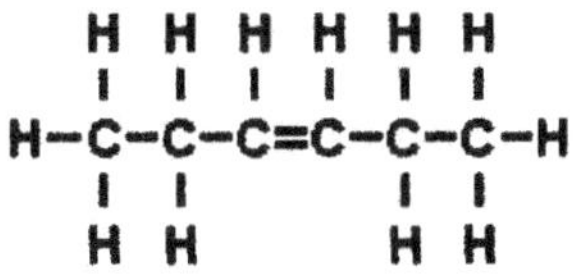

Hexene

Two cyclic aliphatic hydrocarbon structures of the straight chain hexane are shown below in both simplified and written-out forms for cyclohexane. Cyclohexene, which contains a double bond, is also shown and has the suffix **-ene** indicating the double bonds in the structure.

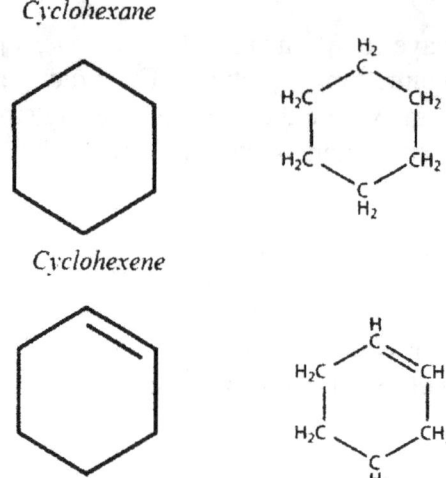

Aromatic Hydrocarbons

The other major category for organic carbon molecular structures is Aromatic Compounds. The denotation as "aromatic" is a chemical classification and does not necessarily mean the compound has a specific "aromatic" odor, although many do.

Aromatic Compounds are cyclically bonded carbon atoms that contain alternating or "conjugated" double bonds. One of the simplest aromatic compounds is Benzene shown below, along with the abbreviated form often utilized by chemists. The aromatic compounds can also be substituted with the other atoms listed above such as oxygen, sulfur and nitrogen. Note too that the aromatic conjugated structure is simplified in another way by use of a circle inside the cyclic carbon structure as indicated for phenol shown farther below.

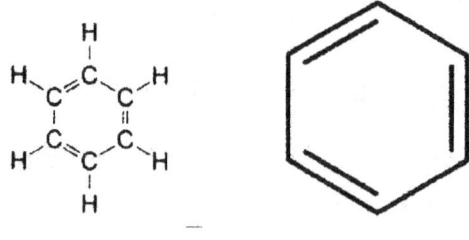

Benzene

Effect of Functional Groups

The odor of chemical compounds can be significantly altered by introduction of functional groups into the molecule. Two functional groups which have a significant impact on the odor of molecules are the **Hydroxyl Group** and the Carbonyl Group. The effect of these functional groups on the characteristic odor of the chemical constituents is described below.

Hydroxyl Group (Alcohols)

As mentioned, other atoms such as Oxygen (O) can be bonded to the chains as shown for a very simple alcohol, Propanol, shown below. One hydrogen is eliminated from the carbon atom to accommodate the bond with the oxygen and another hydrogen is bonded to the oxygen to form a hydroxyl group (-OH), all following our bonding rules above. An aromatic alcohol, phenol, is also shown below. Note the suffix is now **-ol** instead of **-ane** or **-ene**.

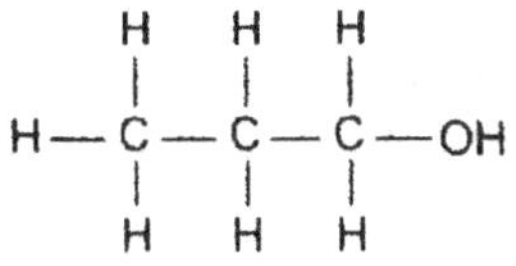

Propanol 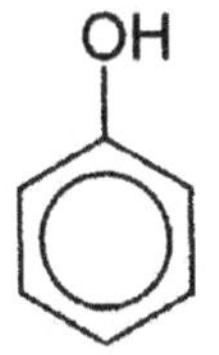   Phenol

The hydroxyl group is the characteristic functional group for all the compounds termed "Alcohols." Alcohols are a very valuable class of compounds for use in perfumery, one of which is Geraniol shown below. Geraniol is also a Terpene which is further discussed below. Since organic molecules can be formed in different ways and contain a variety of functional groups, they can fit into several of the different categories described here and below. By using these designations perfumers can more easily access and assess the value of a particular chemical for a fragrance formulation.

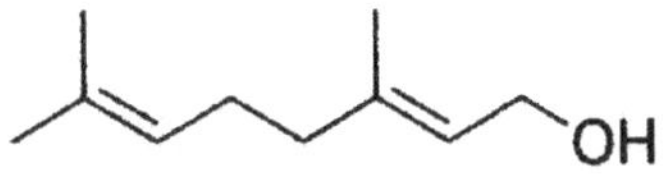

Geraniol (rose-like scent)

Some additional alcohols showing the diversity of structures important to perfumery are (13-16,56):
Cedrol is a natural isolate of Cedarwood Oil and the synthetic analog has even warmer and drier aspects.

Dihydromyrcenol or DHMOL is a synthetic that imparts a powerful, fresh lime like, citrusy-floral sweet odor and is important to the Aquatic type perfumes;

Ethyl Maltol or Veltol is both an aldehyde and an alcohol synthetic compound with an intensely sweet, cotton candy aroma with caramel and chocolate nuances;

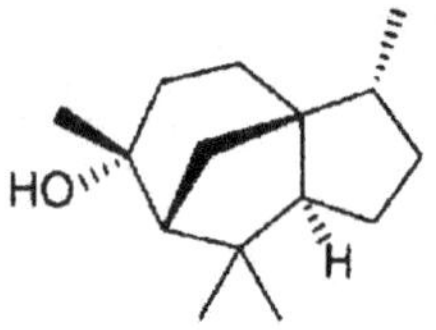

Cedrol 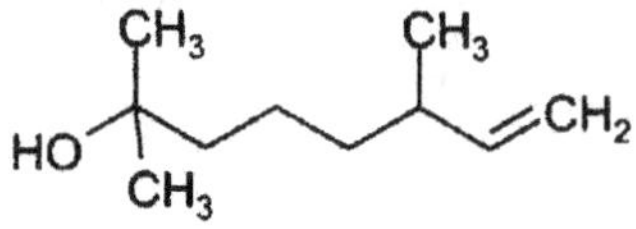   Dihydromyrcenol 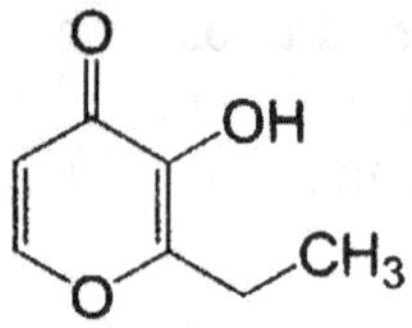   Ethyl maltol

Eugenol is an aromatic alcohol and is natural isolate from clove oil now synthesized for its sweet, warm carnation note;

Farnesol is another natural isolate, a long chain carbon structure, from musk and ambrette. It is synthesized for its sweet, fresh, floral aroma suggestive of Lily-of-the-Valley;

Geraniol is an isolate from Rose, Palma Rosa and Geranium and synthesized for its fresh, sweet, rose-like scent (see structure above);

Linalool is a terpene alcohol, a natural isolate from coriander, with the synthetic compound providing a spicy floral scent. It is also known for its antimicrobial, anti-inflammatory, sedative, and stress-relieving properties.

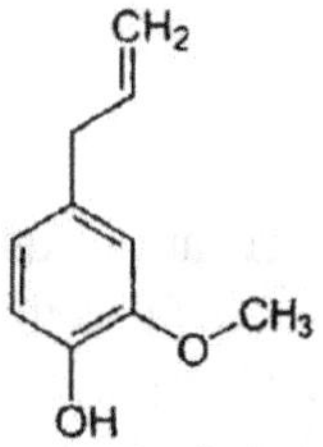

Eugenol 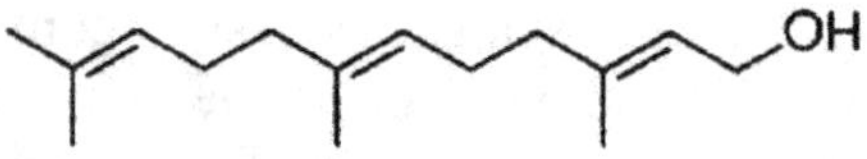   Farnesol 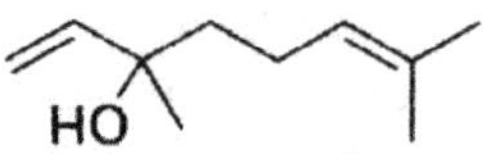   Linalool

Carbonyl Group (Aldehydes & Ketones)

Another functional groups that can be attached to both the aliphatic and aromatic compounds is the Carbonyl Group (-CHO). When a Carbonyl Group is attached at the end of a carbon chain or off a branch point the compound is termed an **Aldehyde;** however, if the carbonyl group is within the carbon chain, bonded to two carbon atoms, it is termed a **Ketone** as shown below. The R and R' represent a carbon atom or carbon chain.

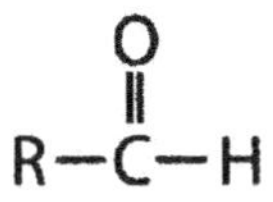

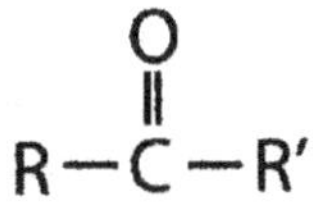

Carbonyl — $-\overset{\overset{O}{\|}}{C}-$

Aldehyde — $R-\overset{\overset{O}{\|}}{C}-H$

Ketone — $R-\overset{\overset{O}{\|}}{C}-R'$

Several simple aliphatic aldehydes are shown below as well as the aromatic aldehyde, Benzaldehyde.

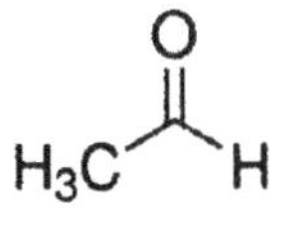

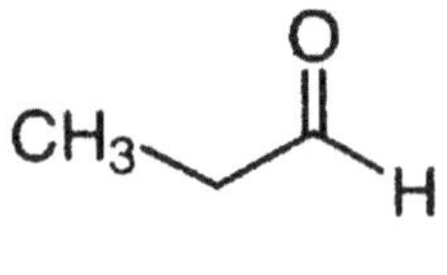

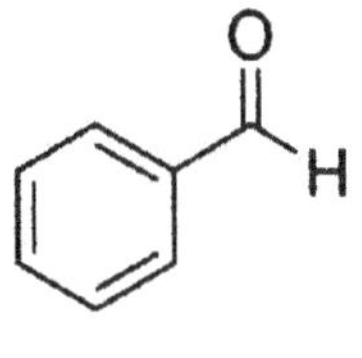

Acetaldehyde   Propionaldehyde   Benzaldehyde

Straight chain aldehydes are very valuable in perfumery (13,16,56). Practically all modern fragrances contain aldehydes, although many have low stability. It has been determined that short carbon chain aldehydes are usually malodorous, but the longer chain aldehydes (C7-C13) have a much more pleasant odor.
The odors for the aldehydes of successively higher numbers of carbons are:
C7 (heptanal) herbal green odor, C8 (octanal) orange, C9 (nonal) rosy,
C10 (decanal) orange rind, C11 (undecanal) clean,
C12 (lauryl aldehyde) lilacs/violets, C13 (tridecanal) waxy/grapefruit.

The structure of C7-heptanal and the longer chain C13-tridecanal are shown below.

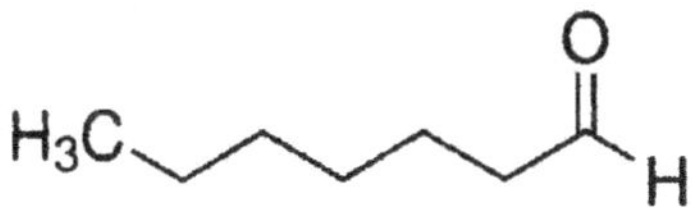

C7-Heptanal   C13-Tridecanal

There are a number of other aldehydes and related lactones (cyclic esters) important to perfumery including:
C12-MNA (methyl undecanal) amber, mossy odor with metallic, waxy nuances,
C14 (γ-undecalactone) referred to as "peach aldehyde' but has a lactone structure,
    peachy, creamy odor notable in *Mitsouko* by Guerlain,
C16 (ethylmethylphenylglycidate) referred to as "strawberry aldehyde" but has a
    more complicated structure, sweet, fruity strawberry odor and
C18 (γ-nonalactone) also a lactone with a creamy, coconut aroma.

Lyral is another synthetic aldehyde with a pronounced floral-muget odor.

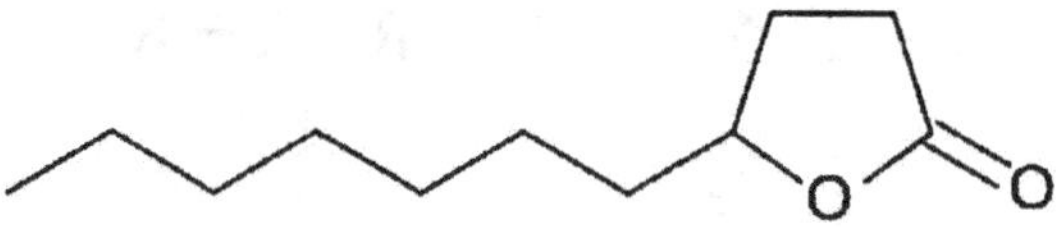

γ-Undecalactone

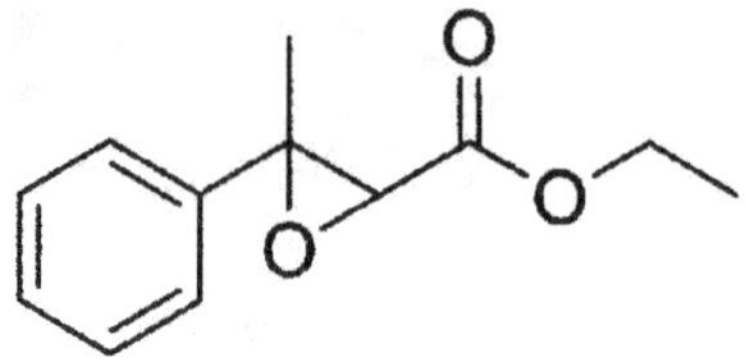

Ethyl methylphenylglycidate

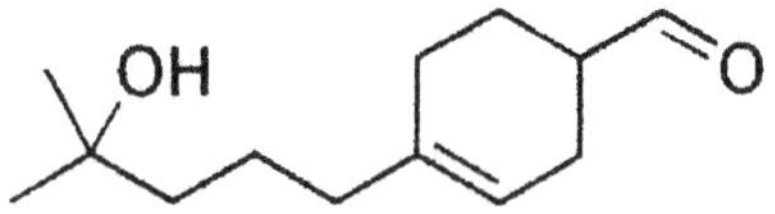

Lyral

Also shown above is γ-undecalactone, which is a **Lactone,** a cyclic ester of a another functional group, a **Carboxylic acid** (-COOH), and also has the suffix **-one** as do ketones. The formation of esters is further described below.

The **Ketones** are bit more chemically complex molecules, with the carbonyl group within the chain structure. They are also important in perfumery. A notable structure in this classification is acetophenone:

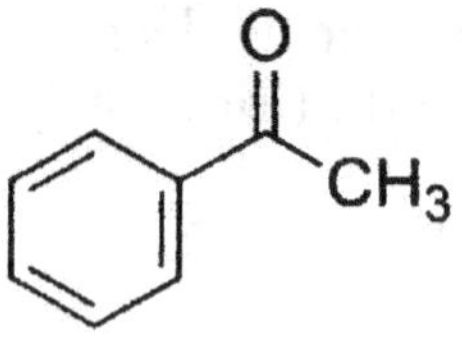

Acetophenone
(pungent hawthorn-mimosa/almond odor)

Some additional ketones showing the diversity of structures important to perfumery include (structures that are not shown are provided with the description of the associated odor in the following section):

Civetone is an isolate of Civet now synthetically available to provide warm sensuality in a perfume, but fecal in large doses (see musks);

Coumarin is the main component of Tonka Bean Oil. It opens with a sweet almond, freshly mown hay scent with a powdery, vanilla undertone. It was employed in perfumes as early as 1882 and is important to both Oriental and Fougère accords;

Damascones provide long-lasting fruity rose notes in perfumes (see rose);

Ionones replicate the aroma of violet and were a major discovery in 1893 (see violet);

Hedione (methyl dihydrojamonate) is a natural isolate from jasmine with a very fresh, crisp citrus odor and important to aquatic accords;

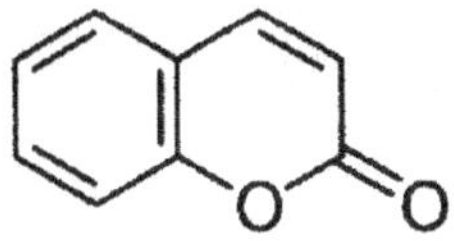

Coumarin 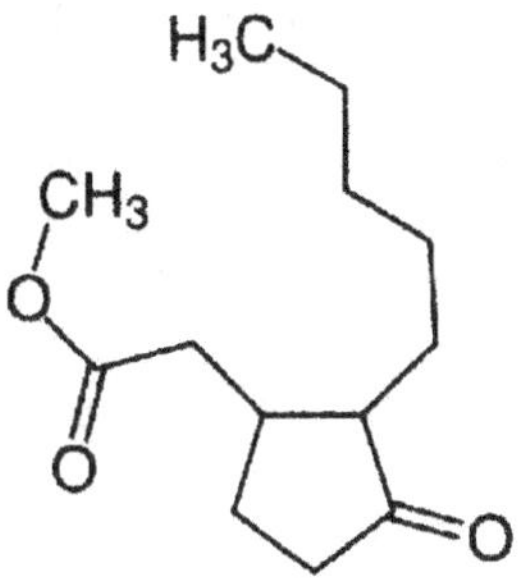   Hedione

Irone is a natural isolate from Iris (Orris) root which is present in such low
 quantities that the synthetic molecule made it possible for general use in
 perfumery;
Calone (Methylbenzodioxepinone) also known as watermelon ketone, provides the
 aroma of the seashore and sea breezes with concomitant marine and ozone
 nuances;
Muscone is a natural isolate from musk now synthetized for use as a soft, sweet and
 sensual note (see musks).

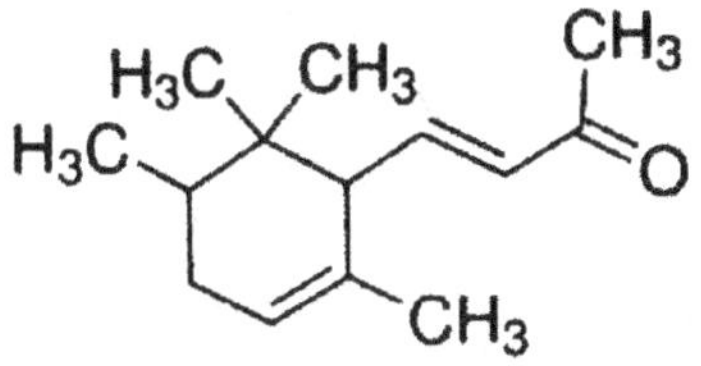

Irone 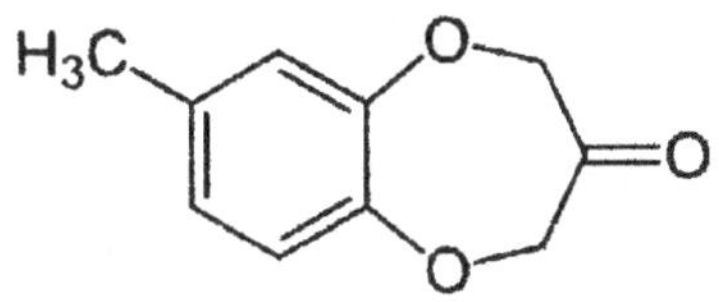   Calone

Esters

Esters are another important class of natural and synthetic chemicals which we referred to above. They are formed from a combination of an alcohol and a second carbon compound, a Carboxylic Acid, that contains another functional group, the **Carboxyl Group (-COOH)**. Esters are known for their intense fruity aromas. The reaction between a carboxylic acid and an alcohol to form the ester, as well as water, is shown below.

$$R-\overset{O}{\underset{\|}{C}}-O-H + H-O-R' \overset{H^+}{\rightleftharpoons} R-\overset{O}{\underset{\|}{C}}-O-R' + H_2O$$

R and R' are carbon chains of the same or different length. The name of the ester is based on the name of the alcohol for the prefix and the name of the carboxylic acid for the suffix, ending with **-ate**. This is shown below for the ester amyl acetate, formed by the reaction of amyl alcohol with acetic acid. The fragrance of a host of esters is summarized in Figure 1.4 and many esters are also utilized for flavoring (27).

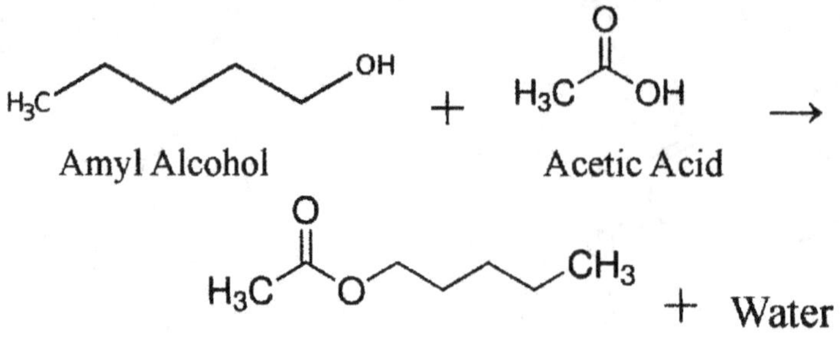

$$\text{Amyl Alcohol} + \text{Acetic Acid} \rightarrow \text{Amyl Acetate (Banana odor)} + \text{Water}$$

The typical fruity/floral odor that results from the combination of the alcohol and the acid is dependent upon the structural characteristics of the two reactants. The odor is related to the relative molecular weight (MW) of the two components of the ester, that is the MW of the alcohol and the MW of the carboxylic acid. The relatively balanced MW of the two components forming **Ethyl acetate**, for example, results in a fruity/brandy like aroma while **Linalyl acetate** (green citrus, lavender odor, see lavender) and **Geranyl acetate** (rose, green odor, see ylang-ylang), from higher molecular weight alcohols, have much less fruity character and their odor more closely corresponds with the alcohols of **Linalool** (floral, citrus orange odor, see ylang-ylang) and **Geraniol** (sweet rose, citrus, see rose) (13-16,56).

This effect is even more noticeable with the higher MW **Phenylethyl acetate** (rose, honey odor) and the phenolic ester, **Paracresyl acetate** (floral-narcissus, animalic odor). However, when the acidic component of the ester is of higher molecular weight, an opposite effect is noted and the acid dominates the odor of the ester with more fruity aspects as noted for Benzoates, Cinnamates, Phenylacetates and Salicylates (13-16).

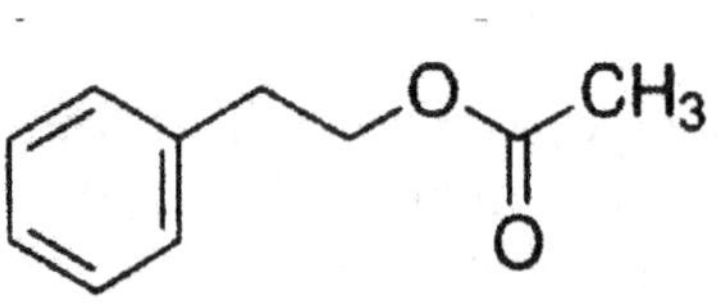

Phenylethyl acetate 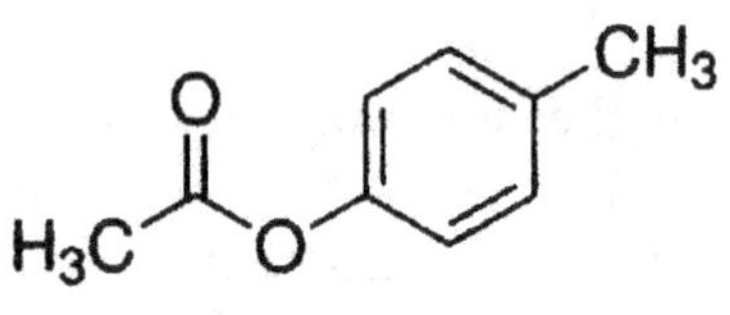   Paracresyl acetate

However, if both parts of the molecule are relatively large the fruity odor is diminished as noted for **Benzyl Salicylate** (fresh azalea & daffodil aromas), **Linalyl cinnamate** (woody, floral, green, herbal), and **Phenylethyl phenylacetate** (heavy, sweet floral, balsamic).

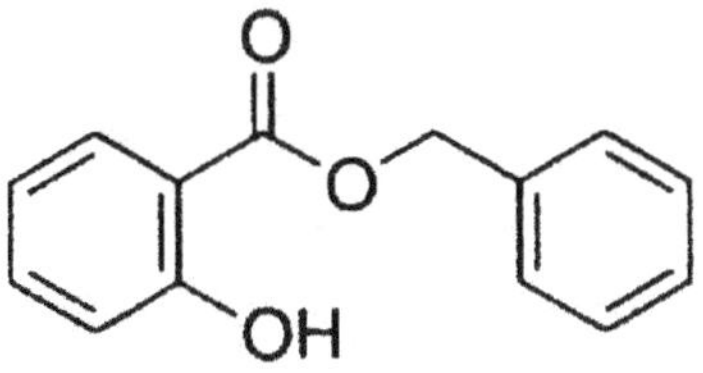

Benzyl Salicylate

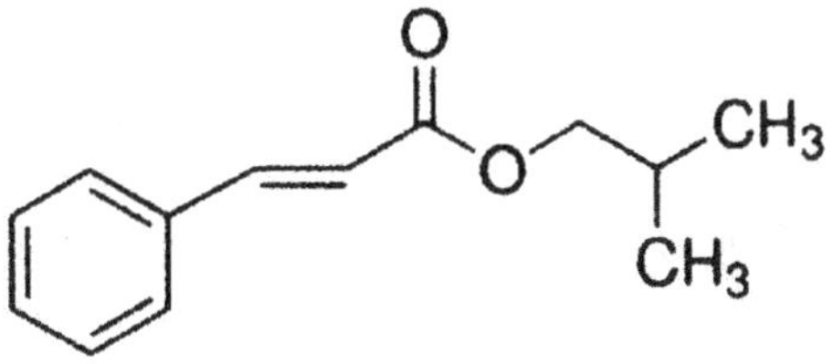

Linalyl cinnamate

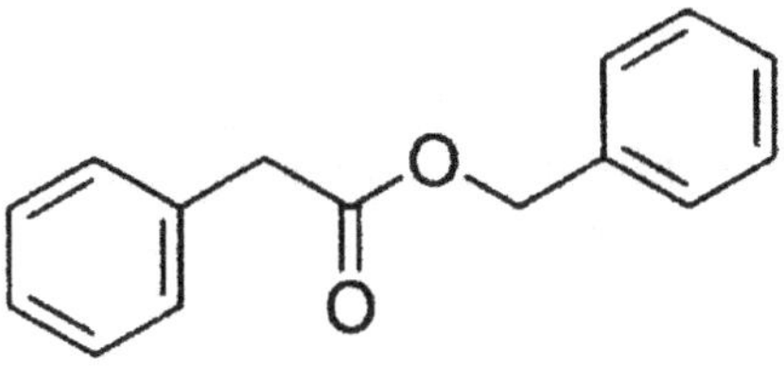

Phenylethyl phenylacetate

Some other important esters for perfumery are **Benzyl Acetate** which is a component of Gardenia and Ylang-Ylang essential oils and produced synthetically for its special fruity, jasmine type aroma and **Ethyl Cinnamate** a synthetic which is also in Storax and has a sweet, oriental odor (13-16,56).

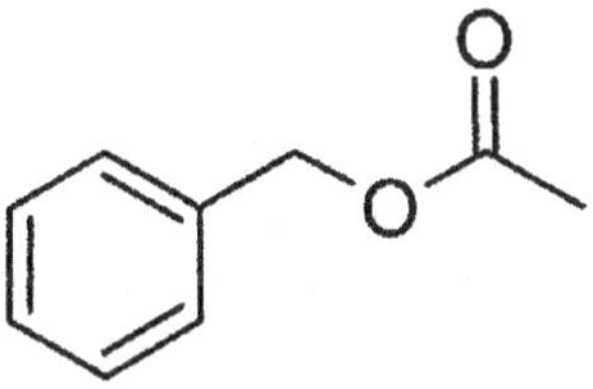

Benzyl Acetate

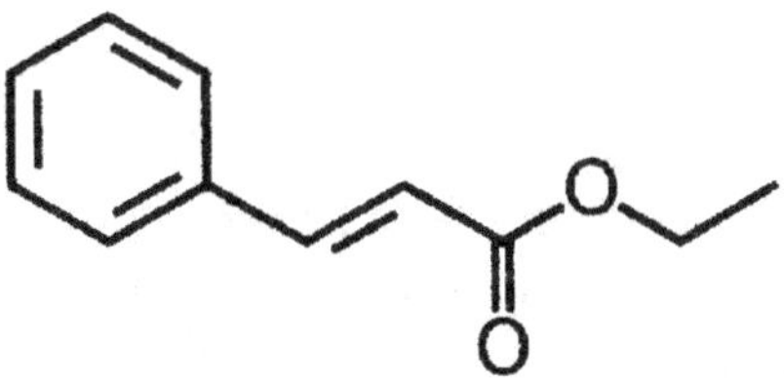

Ethyl Cinnamate

Terpenes

Terpenes are a broad class of compounds and are present in many plant species. They are frequently odorous and useful in both perfumery and aromatherapy (13-16,56). They are aliphatic hydrocarbons uniquely synthesized in nature by building molecules in multiples of five carbon atoms, i.e., Monoterpenes (10 carbon atoms), Sesquiterpenes (15), Diterpenes (20), Sesterterpenes (25) and Triterpenes (30). They can be composed of just carbon and hydrogen atoms or contain any of the functional groups described above. Two natural citrus terpene odorants are Citral (also an aldehyde) and Limonene:

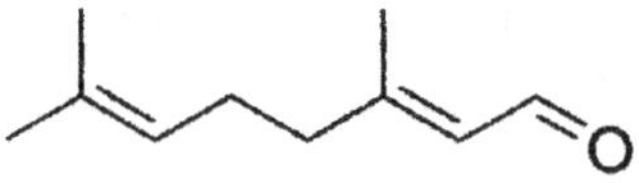

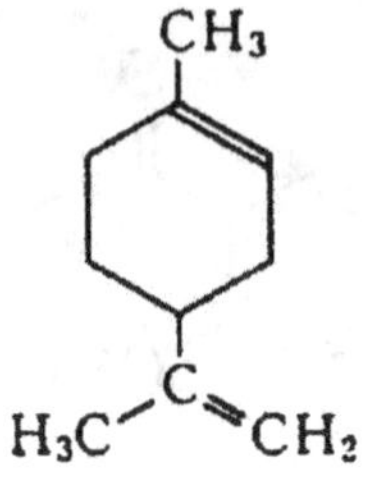

Citral or Geranial (lemon odor)   Limonene (lemon-orange odor)

The five most important terpeniods for perfumery are (13-16,56):
Linalool, Geraniol/Nerol, Citronellol, Citronellal and Citral.
(these floral and citrus odors are further described, with structures, in a susequent section)

Isomers

Isomers are an additional important designation of chemical molecules. Isomers are molecules which have the same atoms of each element, that is, identical formulas but different arrangements of the atoms in space. The distinction is important because the isomers of a molecule may not have the same chemical or physical properties (13-15,56).

There are two categories of isomers. **Constitutional** isomers are different in bonding and connectivity while **Stereoisomers** are different in the spatial, three-dimensional (3D), arrangement of the atoms. For example, fragrant ionones have several constitutional isomers, two of which are shown below, and are very important in perfumery. Note the difference in the location of the double bond in the ring structure for the two isomers. Although the distinction is subtle, the α-ionone has a more floral aroma compared to the β-ionone which is somewhat more fruity/plum with woody nuances.

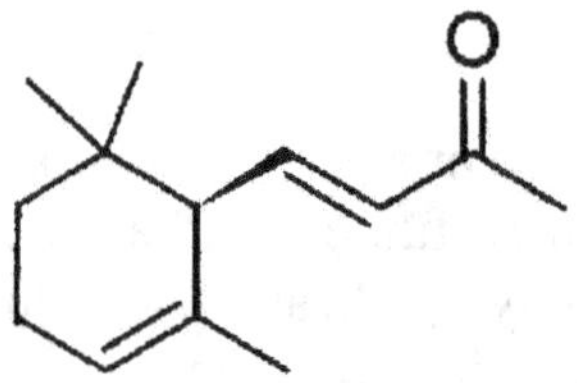

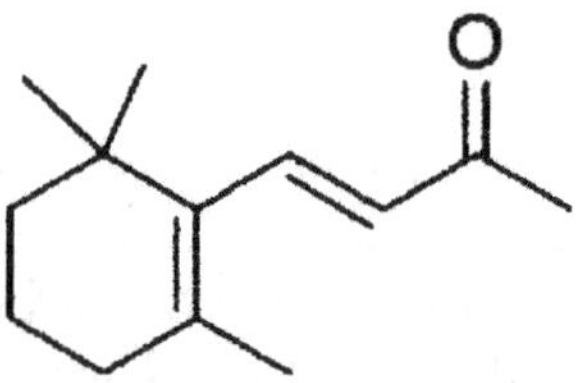

α-ionone β-ionone

Stereoisomers have mirror images and when they cannot be superimposed upon one another, as those shown for carvone in the figure below, the two isomers are referred to as enantiomers. The distinction is important because the R-Carvone has a sweet spearmint odor while the S-Carvone smells of caraway seed oil.

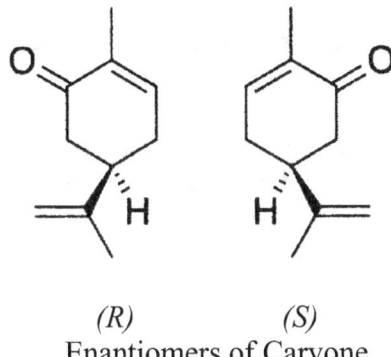

(R)   *(S)*
Enantiomers of Carvone

Summary of Chemical Classification Schemes for Perfumery

The classification schemes used for chemicals, including all fragrance chemicals, is broad and those most important to perfumery have been briefly introduced above and are summarized below. A typical example for each is also provided below and a list of some of the important synthetic compounds in perfumery and their odors is also provided in Appendix I (13-16,56). Again, it is important to note that some of these classification schemes overlap since both the structure and functional groups come into play.

Alcohols/Phenols – cedrol/eugenol
Aldehydes/Ketones – benzaldehyde/hedione
Esters – methyl salicylate
Lactones – γ-undecalactone
Terpene Hydrocarbons – limonene & farnesene
Terpene Alcohols – linalool & farnesol
Terpene Aldehydes/Ketones – citral/β-ionone
Terpene Esters – linalyl acetate
Nitrogen Containing Compounds - indole

Many of the organic molecules discussed in this chapter occur naturally but practically all of them can be synthesized by chemists for use in perfumery and beyond. Often the synthetic chemicals are substantially more inexpensive to create compared to extraction from plant materials. The chemical composition of extracts from some important plant materials to perfumery are further explored in a subsequent section.

Perfume Reactions and Stability

Generally, perfumes are created from a blend of essential oils and synthetics with specific known chemical structures and composition. However, in some cases a perfumer can plan to have several constituents in the blend react after maturation (weeks to months) to form another valuable component in the final formulation.

Probably the best-known use of this method is the formation of what are termed Schiff Bases in the perfume. A Schiff base is formed by the reaction of an aldehyde and an amine compound (contains an -NH_2 group). A typical example of this reaction is between Hydroxycitronellal and Methyl anthranilate to form Aurantiol and water shown below (15).

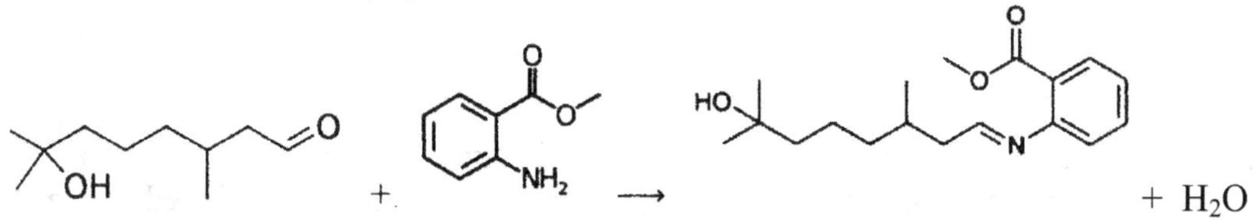

Hydroxycitronellal   Methyl anthranilate   Aurantiol

Aurantiol (Auralva) is often used with floral notes and enhances the top notes of a fragrance. It also has excellent fixative properties. Aurantiol blends well with macrocyclic musks and provides a special oriental character to the fragrance.

Several other aldehydes, in addition to hydroxycitronellal, are used to form Schiff bases in the fragrance formulation including Lyral, Helional, Lilial, Canthoxal and Tripal. It has been found that improved results are obtained if the reaction is carried out *in situ* rather than simply adding the Schiff base to the formulation. The reaction is substantially reduced with final dilution with ethanol. This approach is regularly utilized in formulations of tuberose perfumes (13-15,56).

Another desirable reaction with maturation of perfumes is the formation of hemiacetals. Hemiacetals are formed in a reaction between aldehydes or ketones with alcohols. These reactions are also an important part of the maturation process of perfumes.

There are also a number of adverse reactions that can occur in perfume formulations. Aldehydes and ketones can react by aldol condensation to create undesirable solids in the perfume and acetals can form with the carbonyl compounds eliminating their contribution to the formulation. Oxidation of perfumes can also occur with too much exposure to light and heat.

Gas Chromatography

The advent of gas chromatography (GC) technology in 1961 was a boon to the perfume industry since it allowed the perfumer to take a detailed look at the chemical composition of essential oils and the opportunity to take an informed "peak" at the formulation of classical and competitors' perfumes (23,14,15,56,81).

The GC is a perfect complement to perfumery since it is designed to separate and analyze volatile components of a liquid mixture. The procedure involves injecting a very dilute liquid sample into the injector port of the GC apparatus where it is immediately vaporized and carried by an inert gas to a heated coiled column. The column is packed with absorbent material to which the chemical components of the mixture absorb and desorb based on their differential affinity for the stationary phase in the column. The chemical constituents of the mixture exit the column at different rates and our electronically recorded as individual peaks on a recording chart.

This recording chart is termed a chromatograph and both the height of the peak and the area under the peak are proportional to the amount of material of the specific component. The time for each constituent to emerge from the column is termed the "retention time" and is characteristic for each component based on the flow rate and column specifications, with an additional correction factor (81). A typical gas chromatograph is shown in Figure 4.1 for sweet orange where each peak represents and individual compound in the essential oil.

As the chemicals emit from the exit port they can be directly smelled and identified by an experienced technician or perfumer. However, the attachment of a mass spectrometer to the exit port provides a much more sophisticated method for precise identification of the chemical components of the mixture. Mass spectroscopy (MS) is a more complex process that involves ionization of the molecules by bombarding with electrons which causes considerable fragmentation of the chemical structure into charged fragments. These ions are then separated according to their mass-to-charge ratio and conveyed to a mass analyzer which displays the signals as a characteristic spectrum of each compound. The atoms in the sample are identified by correlation of the fragmentation pattern of known masses to that of the unknown constituent.

The resultant GC-MS chromatograph of a perfume will contain many peaks requiring both a skilled perfumer and technician for interpretation. Many natural materials will show a pattern of constituents (peaks) recognizable by the perfumer, such as sandalwood, ylang-ylang and patchouli. Other materials will exhibit a characteristic peak useful for identification such as styrene for styrax or myristicene for nutmeg oil (56,81).

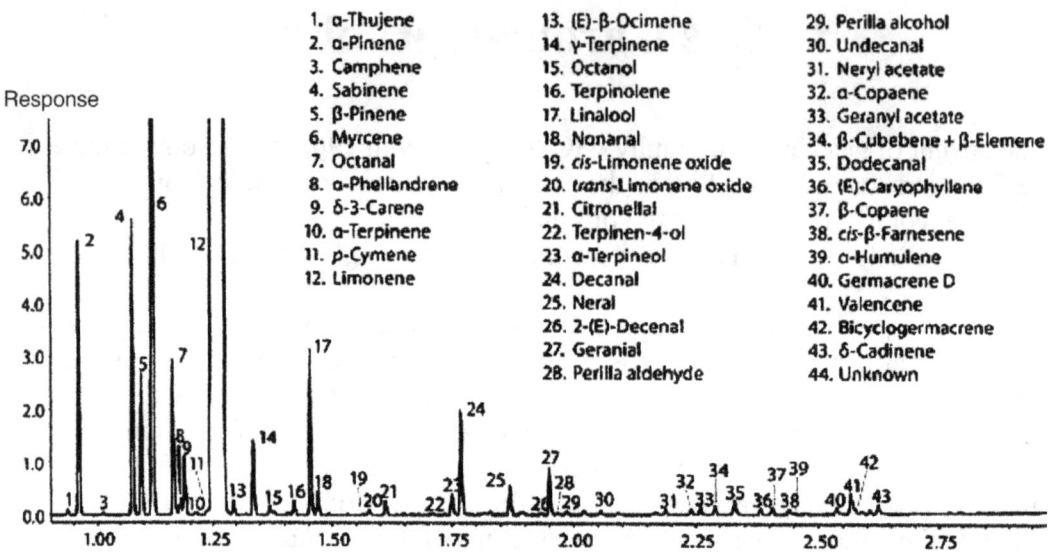

Figure 4.1 Gas chromatogram of sweet orange essential oil. Peak height indicates relative concentration vs retention time of constituent in GC column.

The benefits of GC-MS far outweigh the difficulties with interpretation of the output data, with the method providing rapid approximations of perfume compositions. The perfumer can then use this information for creative modifications and development of unique fragrances.

Another approach utilized by the perfumer for analysis of GC-MS data of a perfume is to look at combinations of constituents indicating the presence of specific bases. The bases have typical relative proportions of the constituents which a perfumer can seize upon for identification. However, difficulties can be encountered when a chemical constituent is part of several different essential oils or it is included in a number of different bases in the perfume. Many of the same terpenes are present in a host of different essential oils (81).

Headspace Technology

A further development of GC-MS technology for the perfumer has been the application of Headspace Technology (14,15,23,81). This approach involves trapping the odor molecules in a receptacle, i.e. an air tight glass bulb, directly from the raw material and transferring to the GC-MS. This is typically employed for analysis of plant blossoms and, of course, it is important to consider the time of day for the headspace collection. This approach gives a much more precise analysis of the odor molecules emanating from the blossom. The complex chromatograph will contain many peaks for the nuances of an aroma that are not obtainable from analysis of an extract from steam distillation, for example.

An important application of headspace technology is for the analysis of flowers which do not have sufficient quantities of essential oil for extraction by common methods or where the extract does not reflect the fragrance of the flower. Muget (lily-of-the-valley) is an example of a blossom which produces very limited odor but headspace GC-MS analysis revealed the major components of benzyl alcohol, citronellol and citronellyl acetate and some minor components of myrcene and rose oxide. The perfumer can then select the appropriate chemicals to create an accord of the flower (81).

The headspace method has also been utilized to analyze fresh tropical fruit. Some of the chemicals identified in the fruit were a range of esters such as methylbutanoate, methylbenzoate and methylhexenoate. Gas chromatography and headspace technology have proven very valuable for discovery of new constituents in natural oils and for creation of new perfume formulations (15,23,81).

Chemical Composition of Odorants

Essential oils are composed of hundreds of organic molecules which provide the complex aroma from the plant materials. Often there are specific chemicals in the mix that provide the prominent odor of an essential oil. This fundamental odor can be reproduced with a combination of synthetic constituents, however some of nuances of the natural material will not be realized from a limited chemical formulation. In this section some important chemical molecules in plant extracts, as well as some synthetic analogs valuable to the perfumer are presented. However, there are many more natural odorants and synthetic materials available to the perfumer. Calkin and Jellinek listed 62 important natural raw materials and 104 synthetics useful in perfumery (15).

Floral Notes and Accords

Floral notes are ubiquitous in both classical and modern perfume formulations. The most commonly utilized florals are rose and jasmine, but also very important in perfumery are lavender, tuberose, ylang-ylang, violet and many other notes discussed in this section.

Rose (*Rosa damascena* & *Rosa centifolia*)

The odor of rose is one of the most widely utilized notes in perfumery. As previously noted, the preferred rose blossoms for obtaining the essential oil are Rose de Mai and Damask Rose. Considerable quantities of petals are necessary to obtain sufficient oil, 4,000 lbs of petals yields only one pound of oil from steam distillation (1 oz = 60,000 roses). Also the cost of picking the petals can be as much as 60% of the price. All of this, of course, results in the very high price of rose oil (8-12).

Rose ottos are obtained by steam distillation and rose absolutes by solvent extraction. The chemical composition of the rose essential oil is not the same from the two different methods of extraction, but the predominant chemical molecules appear in both as shown in the Table 4.1 (13,34). The composition also differs for the odor of the blossom obtained by headspace gas chromatography. The fragrance of rose essential oil is clean, lemony fresh, floral, dark berry and somewhat boozy liquor-like.

Table 4.1 Composition of Rose Essential Oil According to Extraction Method

Constituent	Percent Oil by Steam Distillation	Percent Absolute by Solvent Extraction
Geraniol & Citronellol	63-84	30-34
Nerol	5-10	5-10
Phenylethyl alcohol	1	46-60
Linalool	trace	trace

Reference: Calkin & Jellinek (15)

A more detailed analysis of Damask rose oil revealed over 237 chemical constituents in the oil, with the major ones shown in the Table 4.2. Included is an "Odor Unit" value obtained by dividing the concentration of the constituent with the detection threshold level, in parts per billion (ppb). The odor unit gives an idea of the potency or contribution of the constituent to the odor of the rose oil (13,34).

Significant chemicals in rose oil are the two terpene alcohols of citronellol (38%, rose, sweet citrus, green) and geraniol (14%, sweet rose, fruity) which comprise a substantial amount of the oil. In the 1960s, Professor Ruzicka and Dr. Kovats at ETH in Zurich discovered a series of Rose Ketones as α-, β-, and γ-damascones and β-damascenone in Bulgarian rose oil. They demonstrated the significant contribution that these compounds made to the odor of the rose oil (13).

Note the dramatic impact on the odor of β-damascenone and β-ionone compared to the other components which require much higher concentrations for detection. β-Damascenone, contributes a sweet rose, fruity, plum/grape odor. β-Ionone has a quite similar structure to the rose ketones and has a floral violet, berry-like aroma. The difference between the two is the relative position of the ketone group and the side chain double bond. The ionones are an important contributor to violet odors and are further described in discussion of that flower (82,83).

It was also noted that the α- & β-damascones have odor detection thresholds 10 times higher than β-damascenone. The α-damascone has a more rose-apple odor compared to the β-damascone with a black currant-plum note. Rose oxide is also present in the oil and has an intense rose, metallic aroma. All these constituents have potent odors and have been produced synthetically for commercial use in fragrances. The rose ketones are used to simulate a true rose oil odor quite convincingly (82,83).

Table 4.2 Composition and Odor Threshold of Rose Essential Oil Constituents

Constituent	% of Oil	Threshold Detection (ppb)	Odor Units* Relative %
Citronellol	38%	40	4.3
Gernaniol	14	75	0.8
Nerol	7	300	0.1
Phenylethyl alcohol	2.8	750	0.016
Eugenol methyl ether	2.4	820	0.013
Eugenol	1.2	30	0.18
Farnesol	1.2	20	0.27
Linalool	1.4	6	1.0
(-)-Rose Oxide	0.46	0.5	4.1
(-)-Carvone	0.41	50	0.036
Rose furan	0.16	200	0.003
β-Damascenone	0.14	0.009	70.0
β-Ionone	0.03	0.007	19.2
Paraffins	16		

*Measure of odor potency, values obtained by dividing the concentration of the constituent with the odor threshold in parts per billion (ppb); References: Leffingwell (34), Ohloff (13)

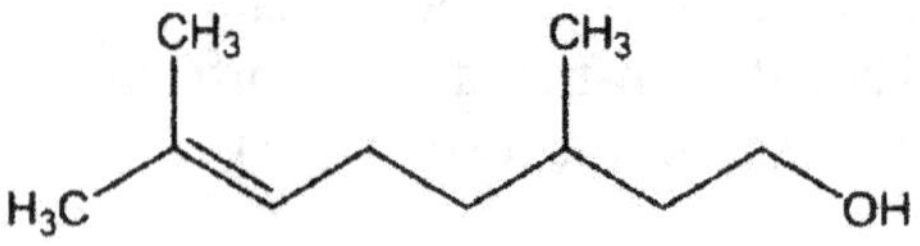

Citronellol

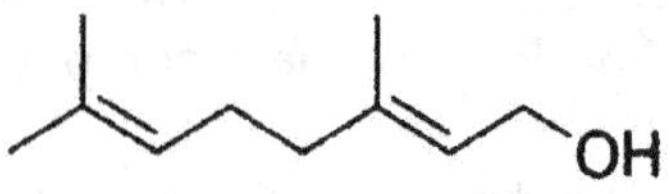

Gernaniol

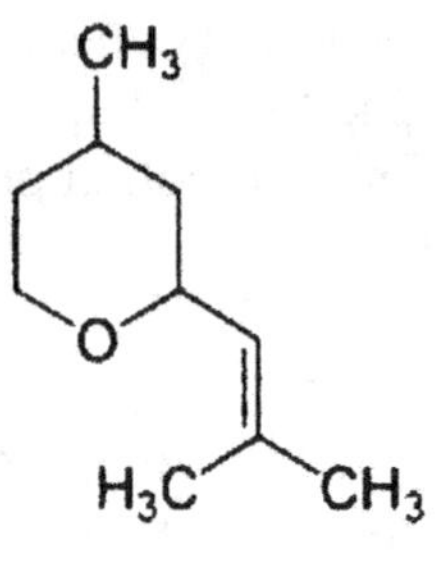

Rose Oxide

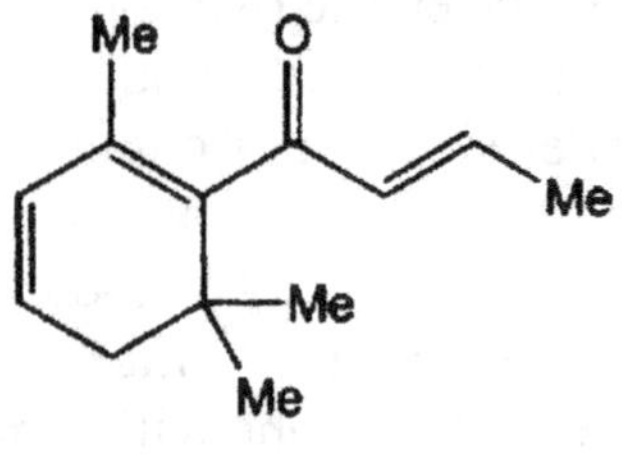

β-Damascenone

The perfume, *Poison* (Dior) has a relatively large concentration of rose ketones in its formulation with 0.4% of α-damascone, 0.09% of β-damascone and 0.09% β-damascenone. *Aqua di Gio (Armani)* contains 0.09% of α-damascone (13).

Jasmine (*Jasminum grandiflorum*)

Jasmine essential oil is very highly esteemed in perfumery similar to rose oils. The flower is very delicate, so the absolute is obtained by solvent extraction. The aroma is narcotic, sweet floral, fruity banana with aspects of apricot and peach as well as a sultry, animalistic smell (8-15). The composition of jasmine oil obtained by enfluerage and extraction by Hesse and Müller is shown in Table 4.3 (15). Both oils contain a high concentration of benzyl acetate while the extracted oil has a much higher concentration of benzyl benzoate.

Table 4.3 Chemical Composition of Jasmine Oil

Constituent	Enfleurage, %*	Extract, %^
Benzyl acetate	60-65	34.0
Benzyl benzoate	-	24.0
Linalool	15.5	8.0
Linalyl acetate	7.5	-
Jasmone	3.0	3.0
Indol	2.5	2.5
Methyl anthranilate	0.5	-

References: *Calkin & Jellinek (15), ^Albert Hesse & Friedrich Müller

Structures are shown below for benzyl acetate with a fruity floral odor, Jasmone as floral, green and Indole with a deep anamalic nuance. The erogenous musky floral odor of jasmine has been attributed by some people to indole, that actually smells fecal at high concentrations. The structure of indole is unique with nitrogen contained in a five-member ring fused with a benzyl moiety. Methyl anthranilate is another nitrogen containing compound and often appears along with indole in essential oils.

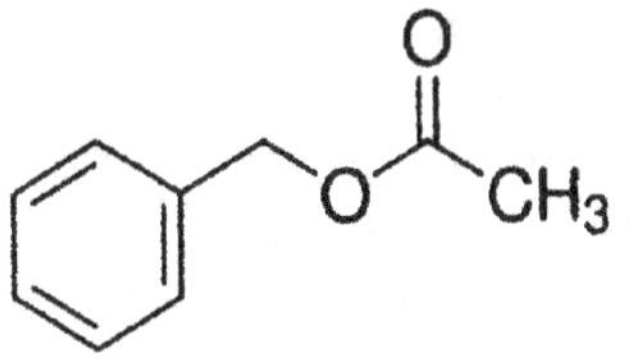

Benzyl acetate 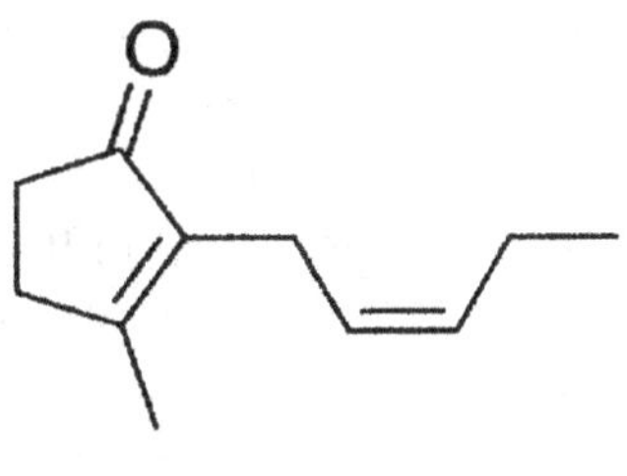   cis-Jasmone 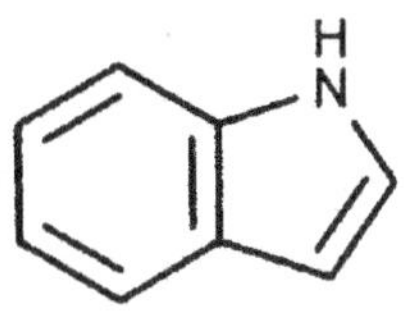   Indole

The analysis of the essential oil composition performed by Hesse and Müller was completed many years ago and modern improvements utilizing GC-MS technology revealed, with a more detailed analysis, over 240 constituents some of which are shown in the Table 4.4 (84-88). There is a very wide range of reported values for the composition of the jasmine oil. The headspace values show a number of additional constituents in significant concentrations, for example, 3-hexenyl acetate and isocaryophyllene. The 3-hexenyl acetate is also referred to as "leaf alcohol" and contributes a green note to the composition. Methyl anthranilate contributes a floral, fruity grape-orange aroma.

Also notable in the table is the low content of indole in both the headspace and the absolute for both jasmine species in the analysis by the recent investigators. However, this may be due to the time of harvest of the flowers. Indole builds up during the night in the cells of the petal tissue and is released when the flower opens during the day. Harvesting the buds before opening results in very low indole content and a different combination of chemical constituents in the extract. *Jasmine sambac* or Arabian jasmine is also known to have low levels of indole and it is used to produce green tea.

As shown in the Table 4.4, all of the extraction results differ from the headspace analysis demonstrating how difficult it is to extract a true jasmine odor. The synthetic, hedione (methyl dihydrojasmonate), with floral green notes and an odor recognition at 15 ppb, is often used to amend jasmine based fragrances, with the additional benefit of adding smoothness and radiance to the perfume.

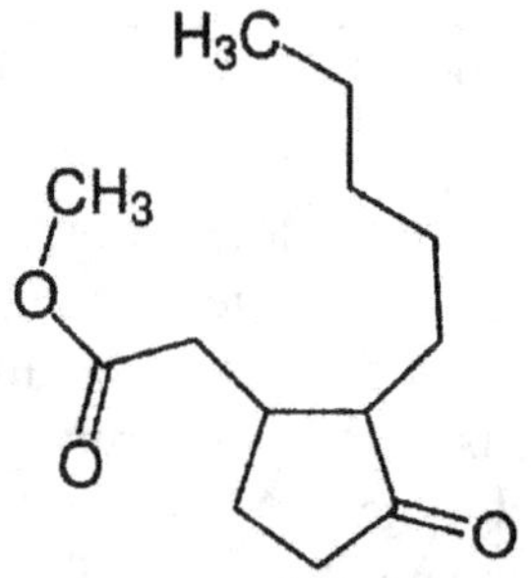

Hedione (methyl dihydrojasmonate)

Ohloff et al (13) reported that jasmine feminine harmony is achieved with a composition of up to 400 parts hedione, 50 parts linalool, 10 parts benzyl acetate, 6 parts cis-jasmone and 1 part indole. Some example fragrances containing Hedione are *Eau Savage (Dior)* at 1.8%, *Calandre (Paco Rabanne)* 7%, *Rive Gauche (Yves St Laurent)* 2% and *Anais Anais (Cacharel)* 8% plus 0.1% jasmine absolute. For further jasmine intensity the synthetic Paradisone, a hedione isomer, has been incorporated into formulations for *Perles de Lalique (Lalique)* and another synthetic jasmine, Magnolione at 13% in *Coriandre (Jean Couturier)*. Another newer, more affordable jasmine substitute, Splendione, has been marketed by Firmenich (13-17).

Table 4.4 Composition of Jasmine Oil According to Extraction Method

Constituent	Headspace Analysis, % J. grandiforum	Headspace Analysis, % J. sambac	Concrete, % J. sambac		Absolute, % J. grand.	Absolute, % J. sambac	CO_2 Ext, % J. sambac
Benzyl acetate	23.7[1]	18.8[2]	4.3[3]	8.2[4]	23.7[5]	10.8[4]	11.6[6]
Linalool	25.0	23.0	13.9	9.7	8.2	12.7	10.2
Methyl linoleate	-	-	-	2.6	2.8	2.2	1.6
Benzyl alcohol	-	8.7	1.3	0.6	1.3	0.9	0.1
α-Farnesene	-	-	18.4	10.3	1.1	11.8	6.3
2-Phenyl ethyl	-	-	2.3	-	trace	-	-
Methyl anthranilate	4.6	0.2	5.5	3.6	1.0	4.1	4.9
3-Hexenyl acetate	13.8	3.4	-	4.2	trace	5.4	3.9
3-Hexenol	-	3.9	0.4	0.3	trace	0.4	0.1
Methyl benzoate	6.3	5.2	2.6	-	20.7	12.7	0.6
Methyl salicylate	2.6	0.9	-	0.6	0.1	0.6	0.6
Indole	-	0.1	14.7	0.2	1.8	0.3	0.9
Isocaryophyllene	13.7	-	-	0.1	-	0.4	0.2
Benzyl benzoate			-	0.6	20.7.	1.0	0.8
cis-Jasmone			0.3	-	1.9	-	0.2
5-Methyl tricosane			-	6.7	-	1.7	1.4
Phytol					10.9	-	-
Hydrocarbons			-	5.5	-	0.6	-

References: [1]Zu et al. (85), [2]Bu et al. (86), [3]Kaiser (87), [4]Rout et al. (84) (avg. of 3 samples), [5]Jirovetz et al. (88), [6]Rout et al. (84) (avg. of 2 samples)

Muget (Lily-of-the-Valley, *Convallaria majalis*)

Lily-of-the-valley is widely distributed in the northern hemisphere and found in dense patches in southern Sweden. It has an elegant and delicate floral fragrance with rose-lemon nuances and a touch of green. The intense fragrance has been used in many perfumes in the past but is somewhat unstable and slightly allergenic.

The above factors combined with the difficulty obtaining suitable amounts of the essential oil from the flowers has resulted in the drafting of many formulas of natural and synthetic ingredients for recreation of the aroma of muget (9-15,89). The earliest (1902) "synthetic muget oil" was a simple combination of 50% geranium oil (geraniol), 30% cinnamic alcohol, 20% farnesol, 2% α-ionone and 1% benzaldehyde (90).

Intensive research to realize a central note of muget odorants led to a number of valuable synthetic substitutes which have found wide-spread usage in perfumery as hyroxycitronellal (1905), cyclamen aldehyde (1919), Lilial (lily aldehyde) (1956), Lyral (1960) and most recently Nympheal (2016). Nympheal has much greater potency and radiance in comparison to Lilial and provides more fullness and creaminess to a fragrance (89,90).

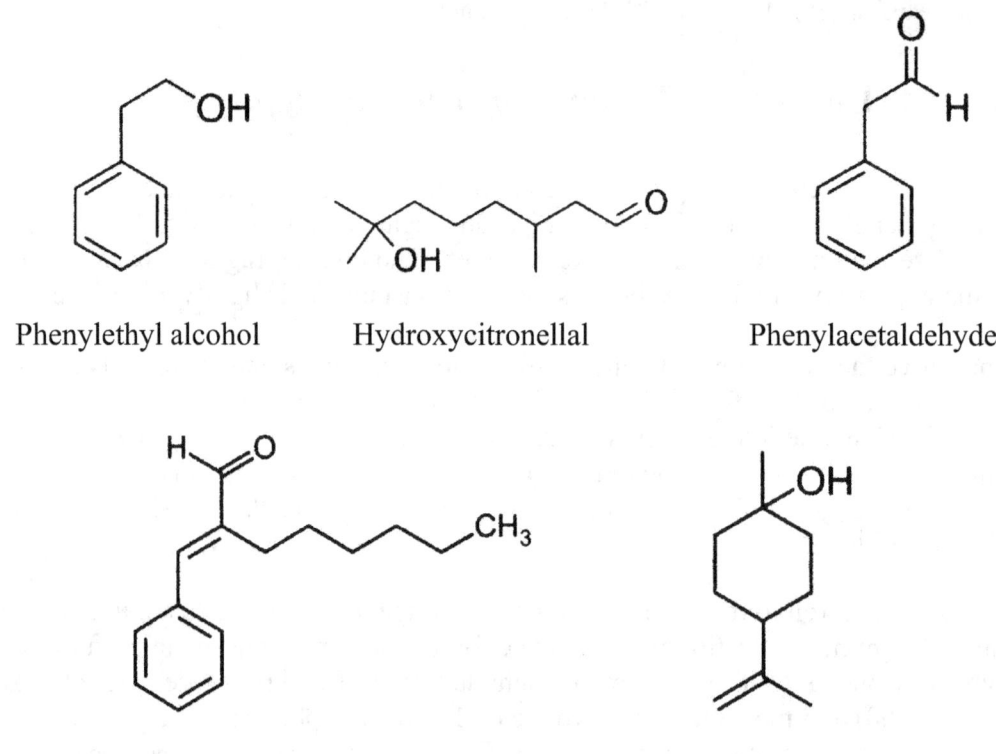

Lilial   Nympheal

A number of fragrances feature lily-of-the-valley notes including the classic *Diorissimo (Dior)* and *Muget du Boheur (Caron)* and more recently *Muget Fleuri (Oriza Legend)* and *Muget des Bois (Coty)* (17).

It has been noted that a variety of floral fragrances, such as jasmine, lilac, muget and hyacinth can be produced from the same eight chemicals mixed in different proportions. Three of the eight chemical structures of Citronellol, Benzyl acetate, and Indole are shown above and the structures for the other five, of Phenylethyl alcohol (sweet floral, mild rose), Hydroxycitronellal (light floral, linden blossom, muget), Phenylacetaldehyde (sweet, green clover, rose & hyacinth), Hexyl cinnamic aldehyde (sweet-oily, floral, herbaceous) and Terpineol (fresh lilac, apple blossom) are shown below (15). Also refer to discussion of muget, lilac and hyacinth that follows.

Phenylethyl alcohol   Hydroxycitronellal   Phenylacetaldehyde

Hexyl cinnamic aldehyde   Terpineol

Lavender (*Lavandula augustifolia*)

Lavender has been utilized for millennia for bath waters and it is a pervasive note in fragrances for men. It has also been highly touted for relaxation and as a sleep aid. The aroma is aromatic, clean, almost medicinal with a eucalyptus/licorice like end. The oil is obtained by steam distillation and contains many terpenoids (9-16).

The **Lavender oil** is obtained from *Lavandula augustifolia* while **Spike lavender** is obtained from *Lavandula latifolia*. There is also a hydrid between the two species, *Lanvandula intermedia*, which yields **Lavandin oil**. The fine lavender oil from *L. augustifolia* is the highly desired option for fragrances, however, by 2009 nearly 90% of the fragrance plants grown were for production of lavandin oil. This oil is less refined than the fine lavender oil with a stronger, medicinal smell. Several environmental and disease factors have affected the growth of *L. augustifolia* and thus the availability of the fine lavender oil (91,92).

The chemical composition varies by species as show in the Table 4.5 (91). As with most essential oils there are over 100 compounds in the extract but many in very low concentrations. Four typical chemical constituents of lavender oil are linalool (spicy floral, see Ylang-Ylang), linalyl acetate (floral, citrus), ocimene (warm, citrus, herbaceous) and camphor (spicy, woody, herbal, like moth balls) (9-13).

Table 4.5 Composition of Lavender Essential Oils

Constituent	Lavender *L. augustifolia*, %	Spike Lavender *L. latifolia*, %	Lavandin *L. intermedia*, %
Linalool	28.9* (25-38)^	49.5*	32* (24-37)^
Linalyl acetate	33.0 (25-45)		26 (25-38)
Eucalyptol		25.9	
Camphor	0.9 (0.0-0.5)	13.0	(6.0-8.5)
β-Caryophyllene	4.6	2.1	
Lavandulyl acetate	4.5		
β-Ocimene isomers	7.5 (5.5-16.0)		(0.5-2.5)
β-Farnesene	2.7		
α-Terpineol	0.9 (0.1-6.0)	1.1	(0.3-5.0)
Borneol		1.4	
Cineole	(0.0-1.0)		(4.0-8.0)
Lavandulol	0.8		
α- & β-Pinene		0.9	
Myrcene	0.5	0.4	

References: *Columns from "Lavender Oil"(en.wikipedia.org); ^Columns of Ranges from Bovil (91)

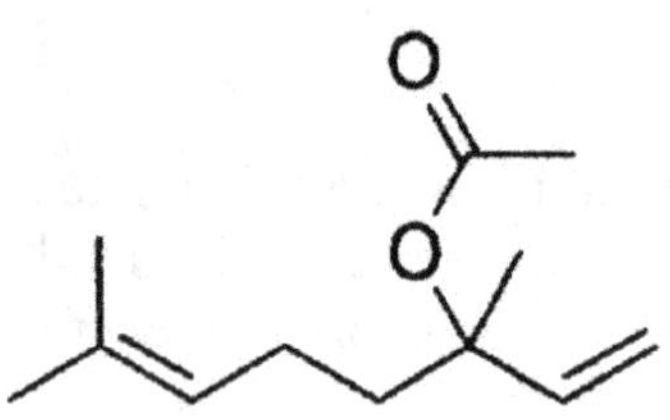

Linalyl acetate 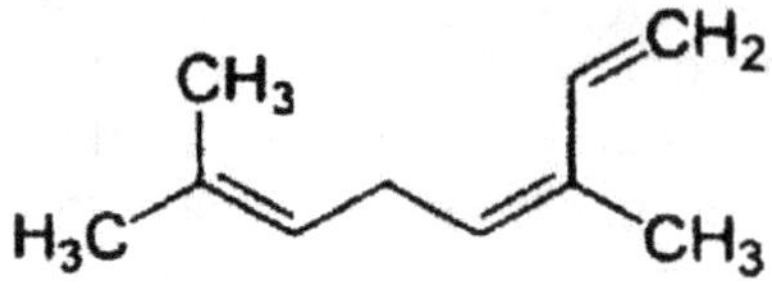   Ocimene 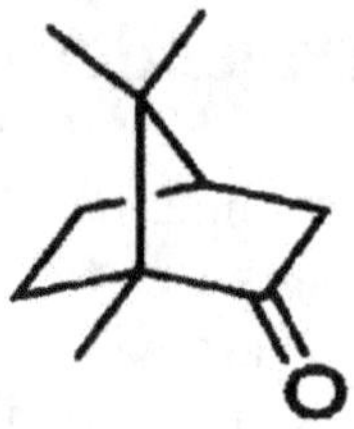   Camphor

A number of fragrances contain various types of natural and synthetic lavender notes such as *Cool Water (Davidoff)* with 3.5% lavandin oil and 9.5% linalyl acetate, *Jicky (Guerlain)* with 33% linalool and *Diorella (Dior)* containing 2% lavandin oil, 11% linalool and 5% linalyl acetate (17).

Tuberose (*Agrave amica*)

Tuberose is derived from the flower of an agave type plant with a flowering spike. The essential oil is obtained by solvent extraction due to the delicate nature of the flowers. Yields are generally low with 3500 pounds of blossoms necessary to produce one pound of oil, making it as expensive or even more than rose oil.

Table 4.6 shows the relative proportions of the variety of compounds in tuberose essential oil from *Polianthes tuberosa* obtained by Rakthaworn et al (93). They reported yields of oil from ethanol extraction via enfleurage at 0.31%, solvent extraction with hexane/ethanol at 0.03% and petroleum ether/ethanol extraction at 0.2%.

Chemical components in the oil include several esters as benzyl benzoate (sweet balsamic), methyl benzoate (medicinal, fruity/floral & powdery) and methyl anthranilate (musky, Concord grape). The power and effect of tuberose oil has been mainly attributed to several lactones including Jasmine lactone also shown below (9-15,93).

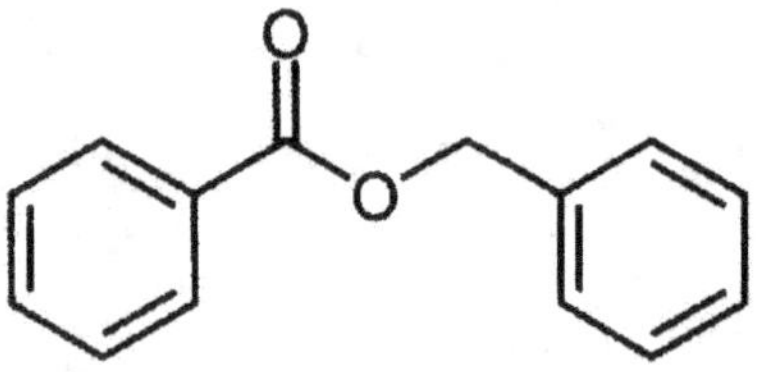

Benzyl benzoate 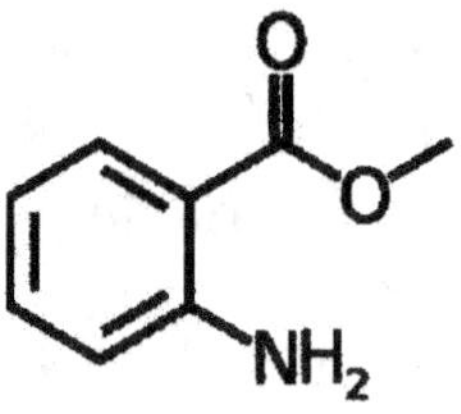   Methyl anthranilate 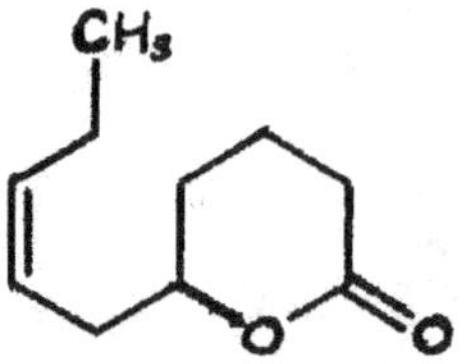   Jasmine lactone

Table 4.6 Composition of Tuberose Essential Oil According to Extraction Method

Constituent	Cold Enfleurage	Hexane Extraction	Petroleum Ether Extraction
Benzyl benzoate	23.6	24.3	10.3
Methyl benzoate	30.2	-	-
Jasmin lactone (7-decen-5-olide)	13.3	15.0	18.1
Methyl eugenol	1.8	1.1	1.9
Methyl isoeugenol	-	2.3	4.9
(E)-Methyl eugenol	8.8	1.9	6.1
Methyl anthranilate	4.5	-	4.2
β-Farnesene	2.4	-	-
α-Farnesol	-	1.9	4.1
Benzyl salicylate	-	1.4	1.3
Methyl salicylate	12.1	-	-
Indole	1.8	-	-
(E)-Citral	1.5	-	-
2,4-dui-tert-butylphenol	-	8.4	-
Hexadecane	-	8.9	1.7
Ecosanol	-	5.0	-
Tricosane	-	2.5	3.7
Pentacosane	-	19.2	29.4
Heptacosane	-	4.6	12.5

Reference: Rakthaworn et al. (93)

Some additional chemicals noted in low quantities in the oil include eugenol, nerol, geraniol and butyric acid, the latter has a sour vomit like odor. The combination of chemicals in the oil result in an aroma that is very exotic, rich, creamy and surprisingly carnal.

Ylang-Ylang (*Cananga odorata*)

The aroma profile of Ylang-Ylang it is sometimes referred to as the "Flower of Flowers" because of the complex nature of the constituents. It is a very valuable scent in the art of perfumery and can be used as top, middle and base notes.
It is a delicate evanescent fragrance and the aroma is rich, floral and sweet with nuances of banana and wood with a dirty touch.

The chemical constituents in the ylang-ylang essential oil include linalool (fresh, floral, citric), geranyl acetate (floral, rose, fruity), caryophyllene (soft spicy, woody, elegant), also benzyl benzoate and benzyl acetate (structures are shown above for the other florals). Caryophyllene has a very complex structure containing only carbon and hydrogen atoms (9-15).

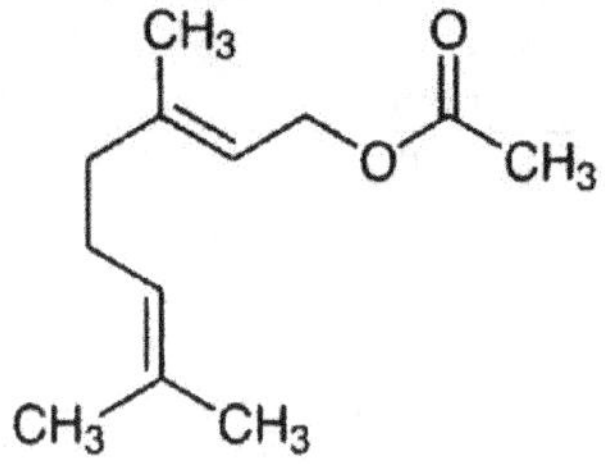

Linalool 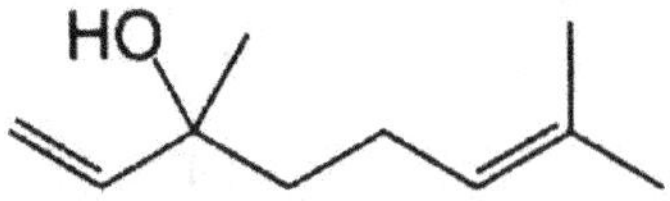   Geranyl acetate 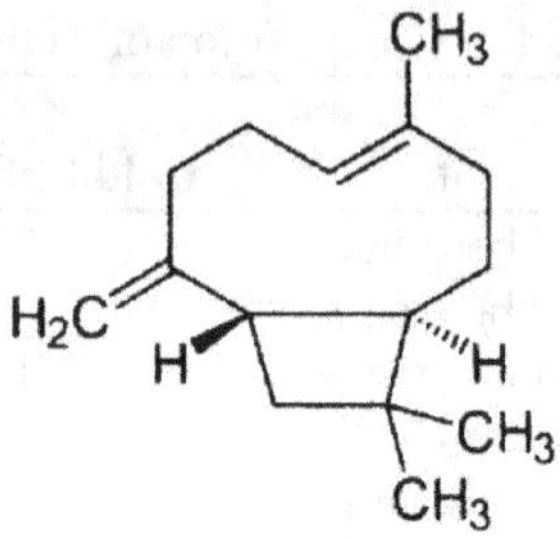   Caryophyllene

Other typical compounds found in ylang-ylang essential oil include germacrene, p-cresyl methyl ether, methyl benzoate, farnesol, farnasene, geraniol, geranial, benzyl acetate, eugenol, methyl chavicol and pinene. The composition of one variant of *Cananga odorata* var. *fruticose* which contains many typical constituents is shown in Table 4.7. However, this variant has a somewhat higher content of caryophyllene and absent are a some of the oxygenated compounds.

Table 4.7 Composition of Ylang-Ylang Essential Oil

Constituent*	Content, %
Linalool	19.0
Geranyl acetate	7.6
Geraniol	1.6
Germacrene D	10.3
Farnesol	1.3
Farnesyl acetate	1.8
Cinnamyl acetate	1.1
Methyl benzoate	3.6
Benzyl acetate	4.6
Benzyl salicylate	1.9
Methyl salicylate	0.2
Benzyl benzoate	7.6
α-Humulene (w/ cadinene)	2.8
β-Caryophyllene	10.7
δ-Cadinene	2.3
Cadinol	1.8
α- & β-Pinene	0.6
p-Methyl anisole	8.4

*Partial composition (*Cananga odorata*), Reference: Gaydou, et al. (94)

Violet (*Viola odorata*)

Violet notes are indispensable to perfumery and are found in almost all fragrance formulations. Both the blossoms and the leaves of *Viola odorata* have been utilized for extraction of violet oil but the low quantities of odor constituents in both and the resulting prohibitive cost of oil production have resulted in a concerted effort to realize synthetic substitutes for this alluring scent (9-12).

Headspace analysis of violet blossoms demonstrated that the primary components, making up 75% of the aroma, are α-ionone (35.7%), β-ionone (21.1%) and dihydro-β-ionone (18.2%) (13,94a). Both α-ionone and β-ionone are efficiently synthesized in industrial quantities and have been extensively utilized in perfumery, with β-ionone the preferred isomer (13-15).

The synthetic Isoraldeine, an isomethyl-α-ionone, has a powdery orris/violet aroma and is in formulations for *Femme (Rochas)* at 15% and *Cuir de Russie (Chanel)* at 6.2%. The formulations for *L'Air du Temps (Nina Ricci)* and *Tresor (Lancome)* have a high concentration of methyl ionones, at 18% (13,15,94a).

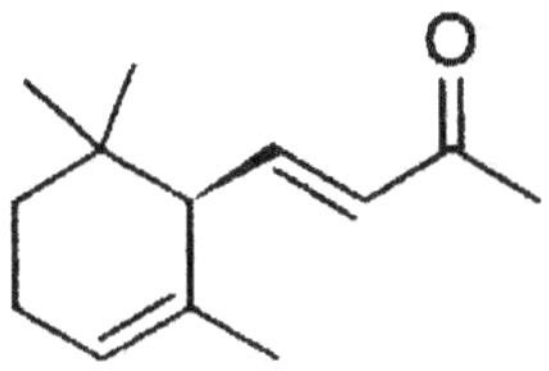

α-ionone 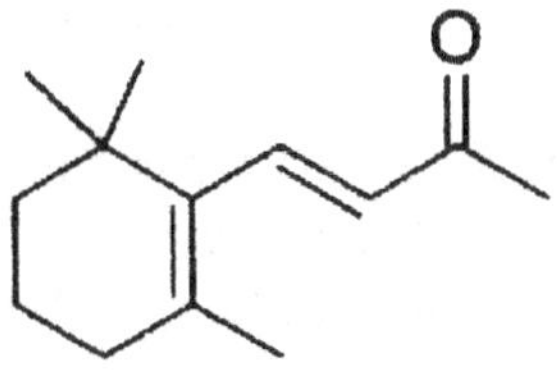 β-ionone 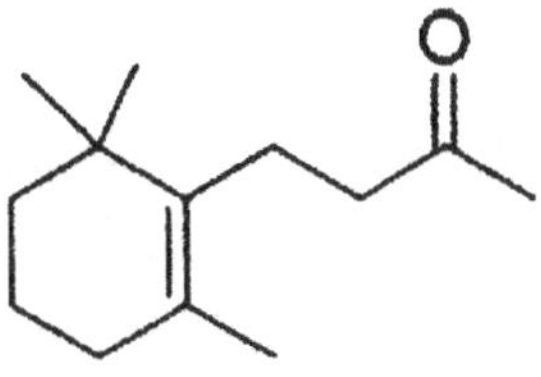   dihydro-β-ionone

Another synthetic violet odorant, Iralia (enriched isoraldeine), was used in the formulations of such classic fragrances as *L'Origan (Coty)* in 1905, *L'Heure Bleue (Guerlain)* in 1912 and more recently in *Poison (Dior)* in 1985. Although in trace quantities, several prominent notes in the violet leaf extract are (Z)-3-hexenal with a grassy odor and 2,6-nonadienol with a strong cucumber aroma (13,15,94a).

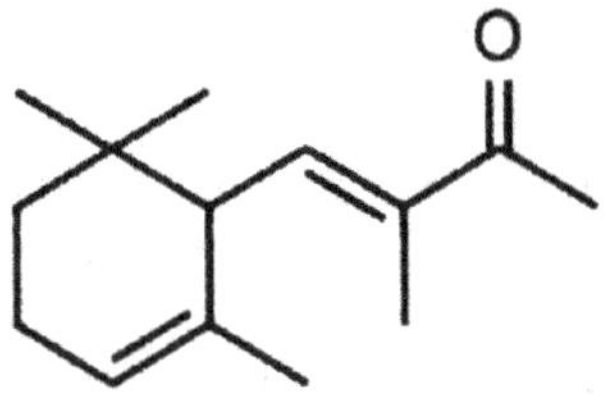

Isoraldeine (isomethyl α-ionone) 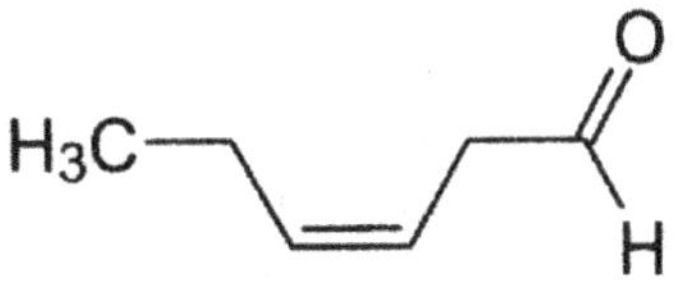   3-Hexenal

Lilac (*Syringa vulganis*)

Originally from the Balkans, varieties of this flowering shrub are now planted in temperate climates around the world. Many of the varieties do not produce a pronounced aroma but *Syringa vulganis* has the characteristic deep rich floral odor hinting of rose with traces of vanilla. The stability of the blossom and content of the oil is insufficient for steam distillation, although small quantities of lilac extract are produced by enfleurage and CO_2 extraction.

The odor of lilac is created with combinations of oils and synthetics. Lilac aroma can be created with notes of rose, lily-of-the-valley, almond and a touch of clove. It can also be reproduced with the combination of eight chemicals listed in the description of Muget. Two notable fragrance chemicals noted in violet headspace analysis are (E)-ocimene (green, tropical foral, vegetable nuances) and benzyl methyl ether (fruity ethereal) (9-15).

(E)-Ocimene   Benzyl methyl ether

Pleasures launched by Estée Lauder in 1995 has a green, lilac floral bouquet and more recent lilac formulations include *Pur Desir de Lilas (Yves Rocher), Lilac (Roja Dove)* and *A Lilac Day (Vilhelm Parfumerie)* (13,17).

Geranium (*Pelargonium sp.*)

Geranium is native to South Africa, but was brought to Europe in 17th century and the first cultivation and distillation of the essential oil was in France. Now a wide range of countries grow and distill the essential oil including France, Spain, Italy, North Africa, Russia, India and the island of Réunion, but Egypt and China dominate the production. The entire aerial part of the geranium plant is subjected to steam distillation for collection of the essential oil and the average yield is 0.15% to 0.2% providing about 70 kg oil per hectare (1-9).

Geranium oil is sometimes referred to as the "male rose" since the floral rose note is nuanced with citrusy green, herbal odors. Whereas rose oil has a sweeter smell, geranium has an aromatic quality similar to lavender and is much less expensive. The oil is an important component of Fougère perfumes and provides a masculine aspect to Chypre compositions.

Roughly 80% of the oil is composed of equal parts of citronellol and geraniol (9-15). Several fragrance formulations containing geranium essential oil include *Dioressence (Dior)* at 5%, *Rive Gauche (Yves St Laurent)* 2.5% and *Cool Water*

(Davidoff) 0.7% (13,17).

Gardenia (*Gardenia jasminoides*)

Gardenia is native to tropical and sub-tropical regions of Africa and southern Asia. The flower is too sensitive for steam distillation and yields are too low from solvent extraction such that synthetic substitutes are utilized for the gardenia fragrance in perfumes.

The aroma of gardenia is said to have aphrodisiac properties and has a pronounced white floral note with green and earthy nuances. The main chemical ingredients from the flower headspace analysis are linalool, methyl benzoate and tiglic acid esters. Methyl benzoate is the main component from enfleurage while (Z)-3-hexenyl tiglate (green, herbaceous, mushroom odor) is the main constituent in the absolute from hexane extraction.

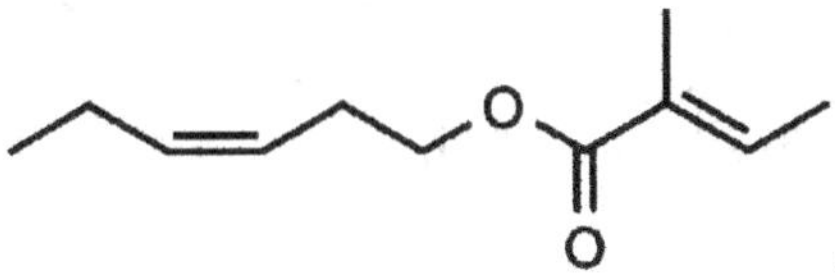

3-hexenyl tiglate

Most synthetic gardenia bases contain styrallyl acetate that has a powerful rhubarb-like green note in combination with jasmine, rose, muget, lilac and a touch of orange blossom. The styrallyl acetate is rather harsh so salicylates and aldehydes (C14 & C18) are added as modifiers. Sometimes the floral "gardenia oxide" (isoamyl benzyl ether) and isovalerate (fruity-citrus) are included in formulations (10-17).

As noted in Table 3.6, Jean-Claude Ellena had a simple formula for a gardenia note comprised of aldehyde C-18 with a fruity/coconut aroma, styrallyl acetate and methyl anthranilate with sweet grape odor. A number of perfume houses have produced gardenia theme fragrances including *Rush (Gucci), Gardenia (Elizabeth Taylor)* and *Gardénia (Chanel)* (17).

Carnation (*Dianthus caryophyllus*)

Carnation is native to the near east and grown in France, Italy, Egypt and Kenya for solvent extraction of the oil. However, as the oil content and yield are extremely small, numerous compounds have been synthesized to replace or supplement the natural oils for perfumery. The main chemical components include eugenol, phenylethyl alcohol, benzyl benzoate, benzyl salicylate, linalool and a host of others. The enhancing softness of the natural oil is related to the plethora of minor components in the composition.

Benzyl isoeugenol with a balsamic note and methyl diantilis with a powdery sweet and smoky odor are two synthetic analogs produced by Givaudan for carnation fragrances. Carnation is the cornerstone of many Oriental type perfumes and is often used to complement a rose note in fragrance formulations. The aroma of carnation is powdery floral with spicy clove attributes (9-15).

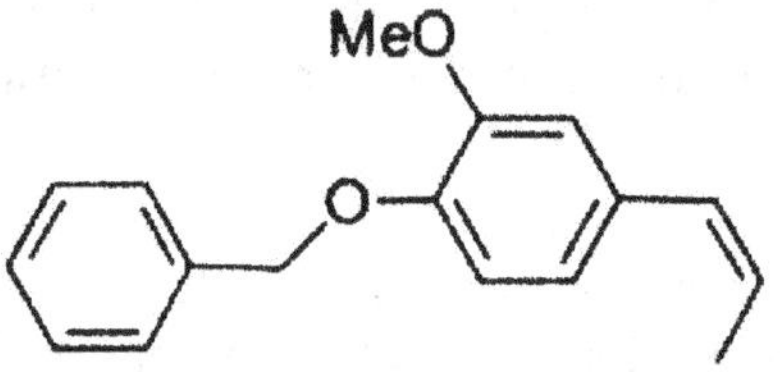

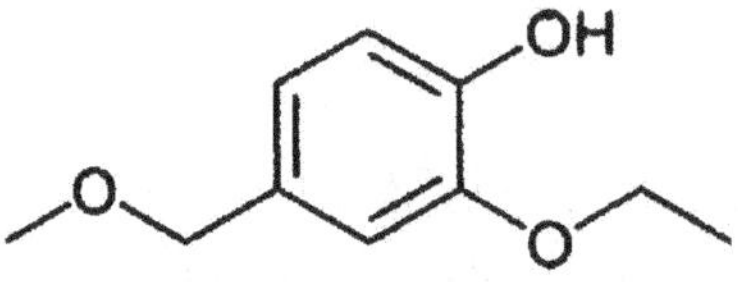

Benzyl isoeugenol   Methyl diantilis

L'Air du Temps by Nina Ricci is probably the most well-known carnation type perfume but there are others containing the alluring scent such *Garofano* launched in 1828 with carnations of the Italian Rivera and more recently *Bellodgia (Caron)* and the Carnation series by Dawn Spencer Hurwitz: *Oeillets Rouges* (15-17). Older (and younger) gentlemen are certainly familiar with the carnation inspired *Old Spice* by Shulton.

Hyacinth (*Hyacinthus orientalis*)

The hyacinth flower is very fragrant but problematical for perfumery since the yield of the natural essential oil is less than 0.02% (9-12). Table 4.8 shows the headspace analysis for the aroma from hyacinth indicating a predominance of benzyl acetate and ocimene (95). Benzyl acetate imparts a pleasant floral, fruity, jasmine type odor while the ocimene provides a green, herbaceous floral character to the hyacinth oils. Ethyl-2-methoxybenzoate also provides a sweet, floral very fruity odor and cinnamic alcohol adds a balsamic aroma. The predominate constituents define just a part of the total aroma from the blossoms and the subtleties and nuances arise from the many minor constituents. The aroma of hyacinth has been described as floral green with vegetal aquatic and spicy nuances.

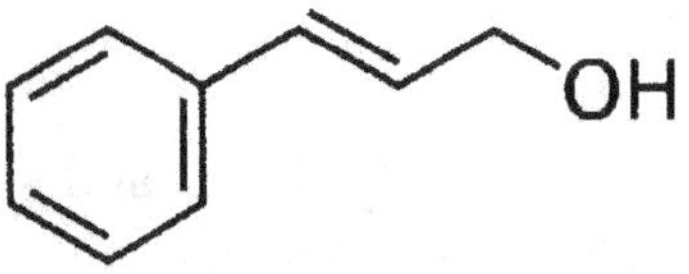

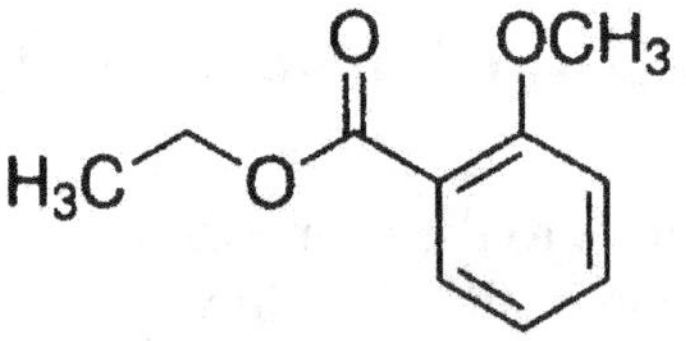

Cinnamic alcohol   Ethyl-2-methoxybenzoate

Several hyacinth fragrances include *Grand Amour* by Annick Goutal and two by Tom Ford, *Ombre de Hyacinth* and *Vent de Fleur* (17).

Table 4.8 Chemical Composition of Hyacinth Headspace

Constituent*	%
Benzyl acetate	36.9-44.2
Benzyl alcohol	1.8-2.0
Benzyl benzoate	2.9-3.1
Cinnamyl alcohol	3.5-4.0
Eugenol	1.6-1.7
Hexenol	0.8-2.7
Hydroquinone dimethyl Ether + α-Farnesene	3.8-3.9
Methyl eugenol	1.1-1.4
Myrcene	1.8-2.1
Ocimene	13.7-15.0
Phenylethyl acetate	1.6-2.5
Phenylethyl alcohol	9.0-16.4
Trimethoxybenzene	1.1-1.3

*Shown are constituents at 1% or greater of 75 total, Reference: Brunke et al. (95)

Orange Blossom & Petitgrain (*Citrus aurantium*)

The essential oil from the orange blossom of the Bitter orange tree, termed Neroli oil, is obtained in a yield of about 0.1% by steam distillation. The petitgrain oil is obtained from the leaves and green twigs of the Bitter orange in higher yields of 0.25-0.5%.

A comparison of the chemical composition of the two oils is shown in Table 4.9 as reported by several investigators (96-98). Neroli oil contains a much larger percentage of linalool compared to the petitgrain oil, where in the latter linalyl acetate is the predominate constituent. Nerolidol and farnesol are also characteristic compounds in the neroli oil.

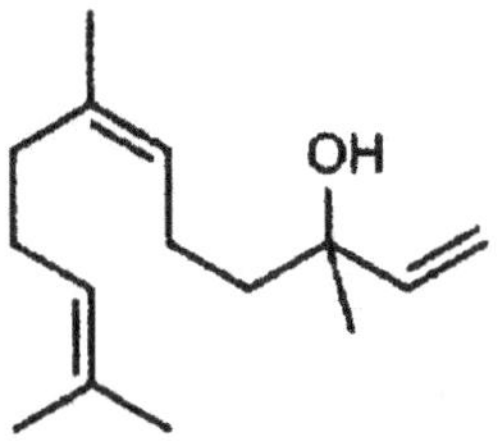

Nerolidol

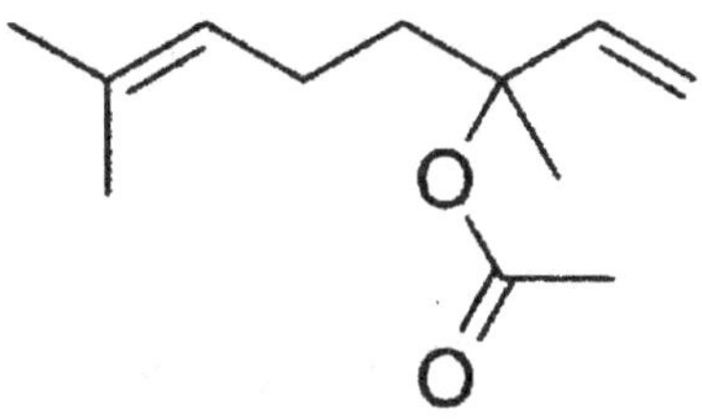

Linalyl acetate

Table 4.9 Composition of Neroli and Petitgrain Oils, %

Constituent	Neroli 1*	Neroli 2^	Petitgrain 1#	Petitgrain 2#
Linalool	44.6	36.0	27.8	24.8
Linalyl acetate	11.9	6.0	54.6	50.1
Limonene	9.3	17.0	1.9	2.5
α-Terpineol	5.6	4.9	3.0	6.2
Ocimene	4.9	6.5	2.0	2.2
Geraniol	3.2	-	-	1.2
Geranyl acetate	3.1	-	2.8	3.4
α-Pinene	3.4	7.3	0.1	0.1
Myrcene	1.5	-	1.2	1.8
Nerolidol	1.9	3.0		
(E,E)-Farnesol	1.5	1.0		
Methyl anthranilate	0.1	0.1		

References: *Bonacoorsi et al. (96) avg of four samples, ^Ohloff et al. (13) & cited sources, #Dugo et al. (98), Mondello et al. (97)

The two oils have a similar but distinctly different aroma. The odor of neroli oil is a sweeter and stronger floral with warm nuances of dried hay compared to petitgrain oil. However, the price of petitgrain oil is one-tenth neroli oil and is sometimes referred to as "poor man's neroli."

Orris Root (*Iris pallida*)

Orris root is the dried rhizome of sweet iris, *Iris pallida*, cultivated in the region of Florence, Italy. The root is aged for some years which results in oxidation of triperpenoid compounds and the development of an exquisite violet odor. The yield of oil is only 0.2% and the prohibitive cost means it is only rarely utilized. The main constituents of the orris root oil are α-irone (20-30%) and γ-irone (30-40%) (9-12,13).

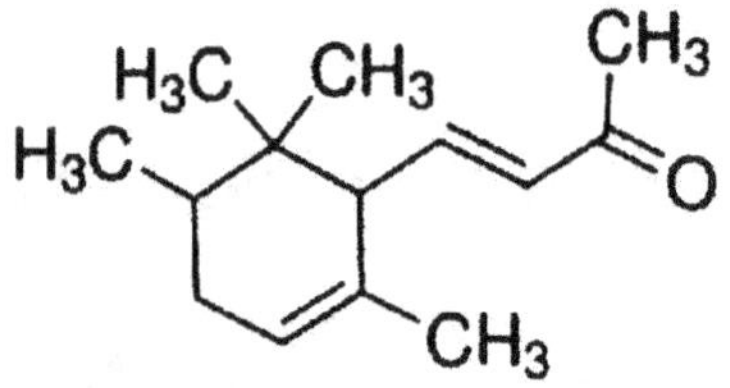

cis-α-Irone

Orris notes are contained in fragrances such as *Chanel No. 19 (Chanel), Royal Orris (Vertus)* and *Orris Noir (Ormonde Jayne)* (17).

Citrus Oil

Citrus oils are extensively utilized in the top notes of perfumes and are an important component of many masculine fragrances. The oils are obtained by cold pressing of the peelings of a variety of citrus fruits, mainly orange, lemon, lime and bergamot. The oil glands in the peel are burst to release the fragrant odor.

A predominate chemical in most citrus oils is limonene at over 90% in orange oil, 40% of lemon oil and 37% of bergamot oil. Lemon oil also contains a high percentage of β-pinene (25%) while bergamot has more linalyl acetate (30%). All three oils have smaller amounts of linalool in their compositions.

As with many plants, the aroma from bergamot oil can vary with the harvest season and is related to the relative composition of the essential oil. The odor in October is floral due to a high content of linalool but with a freshness due to small quantities of ci-5-hexenol in the oil. In February, there is less linalool and more of the fresh smelling linalyl acetate providing a distinctive late odor to the oil (58,59).

The distinctive aroma of orange peel is from the sesquiterpene aldehyde, α-sinensal. The chemical structure for limonene, linalool and linalyl acetate have already been shown above and the terpene, beta-pinene, and α-sinensal are shown here.

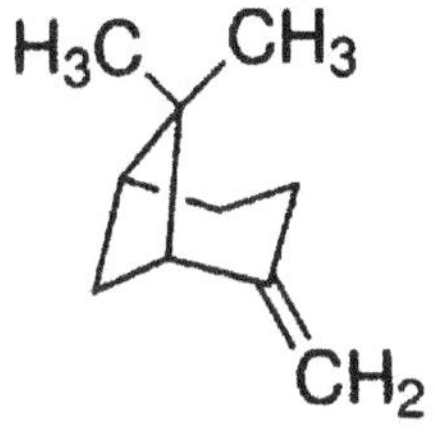

β-Pinene   (E)-α-Sinensal

Some fragrances with characteristic citrus aromas are (17),
Orange: *Eau d'Orange Verte (Hermès), Orange Sanguine (Atelier Cologne),*
Lemon: *Acqua Viva (Profumum Roma), Eau de Fleurs de Cedrat (Guerlain),* and
 Mediterraneo (Carthusia),
Lime: *Aqua Allegoria Limon Verde (Guerlain), Extract of Limes (Geo Trumper),*
Bergamot: *Bergamote Soleil (Atelier Cologne), Velvet Bergamot (Dolce & Gabbana), Aqua Allegoria Bergamote (Guerlain),*
Grapefruit: *Pamplelune (Guerlain), Roma (Laura Bioiotti).*

Fruity and Gourmand Notes

Perfumes with fruity and gourmand notes have been increasing in popularity since the 90s, particularly with the younger generation. Since fruity notes are generally not extractable for use in perfumes, the fragrances rely heavily on synthetic ingredients. A number of synthetics which have a fruity odor are shown in Table 4.10 and more fruity esters are shown in Figure 1.4. There are some natural gourmand notes available such as vanilla, tonka bean, honey, benzoin and coffee, but often the gourmands are obtained from bases or accords. A few gourmand synthetics are shown in Table 4.11 (also in Appendix I).

Table 4.10 Fruity Odors from Synthetic Compounds

Fruit Odor	Chemical Compound
Apple	α-Damascone, Amyl isovalerate
Banana	iso-Amyl butyrate
	Benzyl proprionate (banana-pear)
Blackberry	Cyclopentadecanolide
Coconut	Nonalactone
Citrus	Terpenyl acetate, Limonene
	α-Terpineol, Decanal
Grape	Methyl anthranilate
Lime	Dipentene
Orange	Linalyl acetate, limonene,
	Dimethyl anthranilate
Pear	iso-Amyl acetate, Benzyl acetate
Peach	γ-Undecalactone
Pineapple	Ally cyclohexyl propionate,
	Ally hexoate (Ally caproate),
	Allyl phenoxyacetate, Ally ionone
Prune	Dimethyl benzylcarbinylbutyrate
Raspberry	p-Methoxyphenylbutanone
Strawberry	Strawberry Aldehyde C16
	(Ethyl methylphenyl glycidate)
Watermelon	2,6-Dimethylhept-5-hepenal
Fruity (General)	Phenylethyl isobutyrate (fruity-floral),
	Benzyl isobutyrate (fruity-floral),
	Benzyl isovalerate (apple-pineapple),
	Verdyl propionate (fruity-herbaceous)

References: Curtis & Williams (43), Appell (41, 42), Rouditza (63-66)

Table 4.11 Synthetic Gourmand Compounds

Compound	Aroma
2,3-Dimethyl pyrazine	Nutty, cocoa
Ethyl maltol	Cotton candy, caramel
Ethyl vanillin	Creamy vanilla, caramellic
Coumarin	Balsam, sweet almond, mown hay powdery, vanilla undertone
Methyl acetophenone	Creamy, vanilla, cherry
Methyl cinnamate	Cinnamon, strawberry
Methyl-5-furfural	Carmel-maple, nutty
Prenyl benzoate	Balsamic chocolate, vanilla

References: The Good Scent Co. (12); (creatingperfume.com)

Important Base Odorants

Base notes often make up a substantial proportion of perfume formulations. They are also critical fixatives that provide the long-lasting effects of the fragrance.

Patchouli *(Pogostemon cablin)*

Patchouli oil has a connotation with the "hippies" of the 1960s who used the oil for the earthy aroma. The oil is obtained by distillation of fermented leaves of the plant and 72 constituents have been identified in the mix. Patchouli is often used as a base note and fixative and has a sweet, musky and earthy aroma. Patchouli alcohol (earthy, camphor, powdery) is a major component of the essential oil, up to almost 40%, with α-bulnesene (14%, woody), α-guaiene (11%, sweet, woody, balsam, peppery) and other components (99). Patchouli is pervasive in perfumes and several examples were given in the discussion of patchouli in the chapter on essential oils (99).

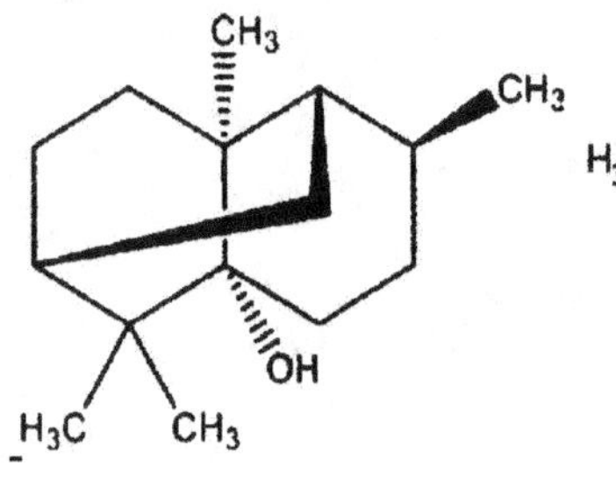

Patchouli alcohol 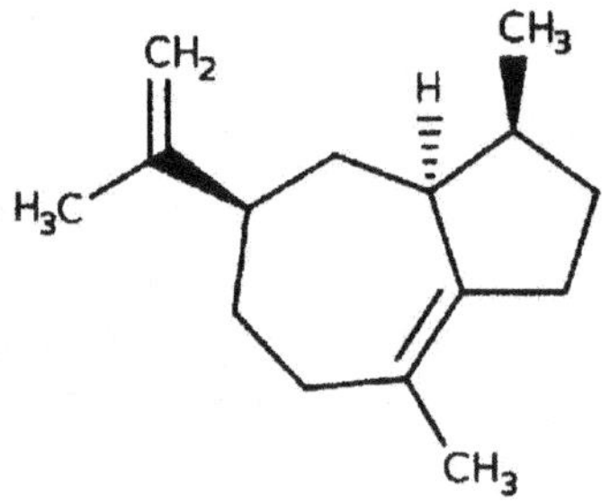 α-Bulnesene 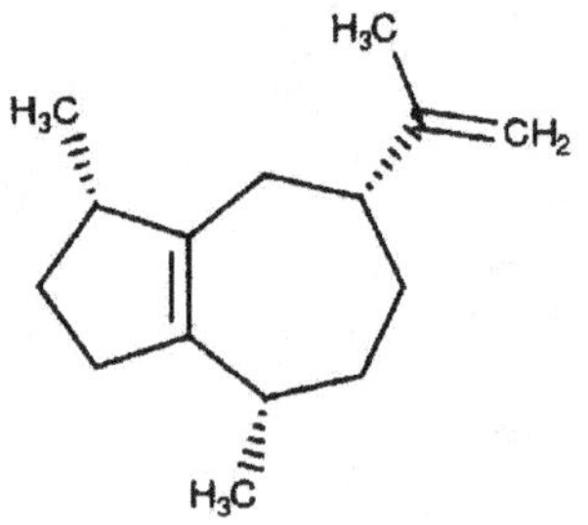 α-Guaiene

Vetiver (*Chrysopogon zizanioides*)

Vetiver, like patchouli, is an important base note and fixative for many fragrances and is used to create Chypre accords. The roots of the grass are extracted to collect the oil. The oil is a complex mixture of over 150 sesquiterpenoid constituents with the composition and quality highly dependent on the origin of the oil. The aroma of vetiver oil is earthy, woody, leathery and smokey sometimes with grapefruit and rhubarb nuances. The major constituents are khusimol (25%) and α-vetivone (9%) (9-12,100).

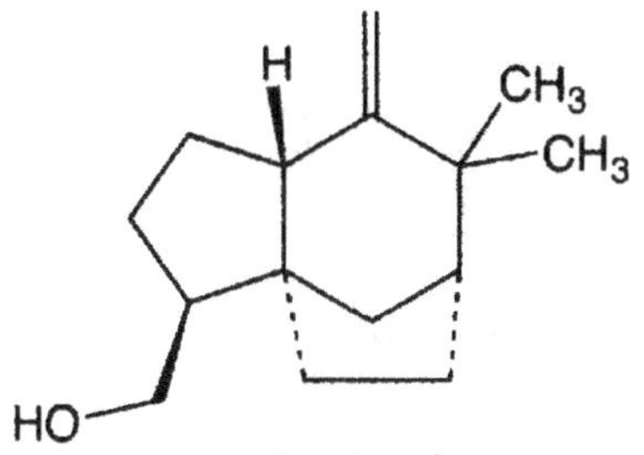

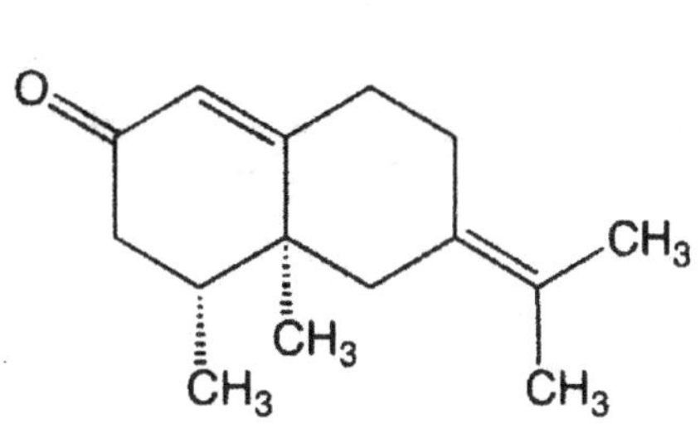

Khusimol   α-Vetivone

It has been reported that khusimol has a woody, earthy aroma; however it may be that the primary source of the odor of vetiver is from diastereoisomeric derivatives of khusimol (13).

Sandalwood (*Santalum album*)

The sandalwood oil is typically obtained by steam distillation of the ground wood and the yield varies with the age and location of the tree. The highly desirable aroma is existent in very many fragrances as a base note and fixative. The odor is soft, woody, milky with a mild green note and a lingering scent. The chemical composition is primarily of the two isomers of santalols, with about 50% α-Santalol and 20% β-Santalol (9-12).

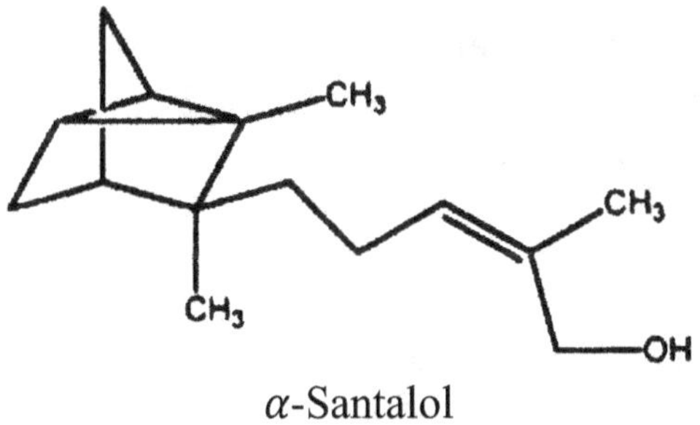

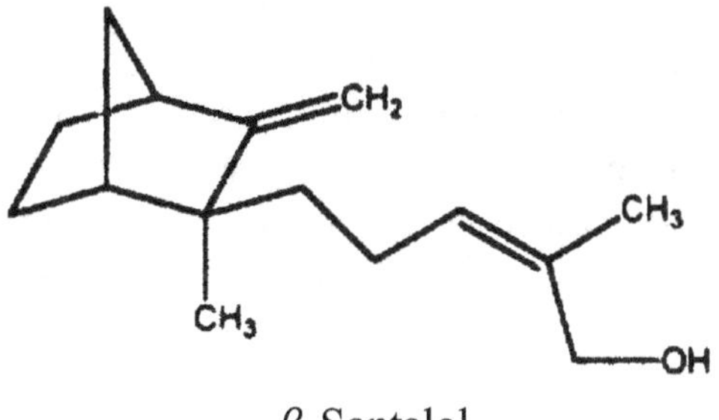

α-Santalol   β-Santalol

However, with the loss of the sandalwood forests around the world and the exorbitant price of the oil, perfume chemists have pursued a synthetic substitute for many years. The β-Santalol constitutes 25% of the essential oil and is the molecule primarily responsible for the pleasant aroma (13-15,26,101).

Hundreds of synthetic analogs of β-Santalol were produced by replacing and modifying various segments of the molecule, but many gave off quite different aromas. With continued hard work and a bit of serendipity a number of synthetic analogs are now available for use in perfumery. Santaliff was produced by IFF and Javanol by Givaudan. Javanol has a natural, creamy sandalwood note with nuances of rose. Also note the unique structure of Osyrol below which also imparts a sandalwood aroma (15,26).

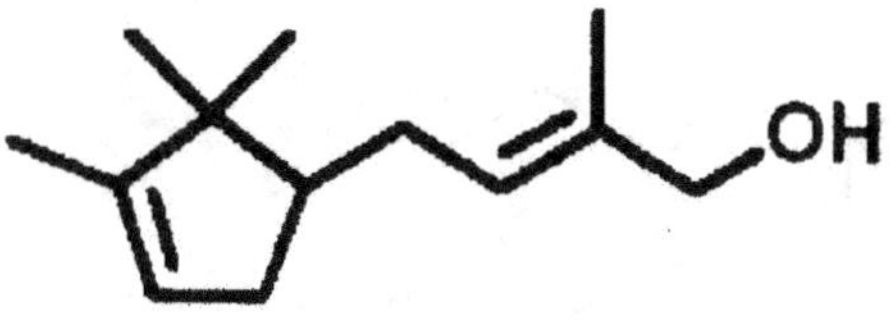

Santaliff

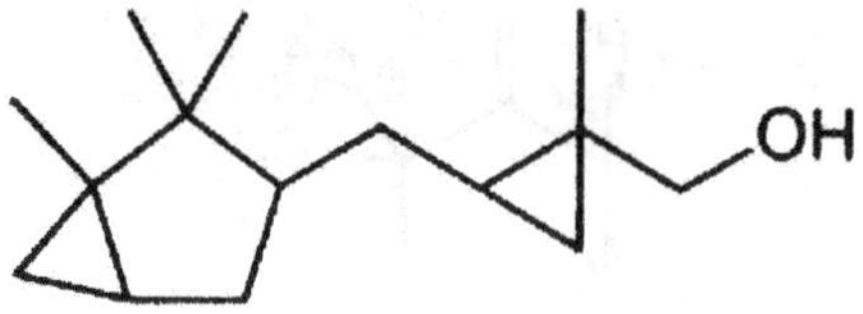

Javanol

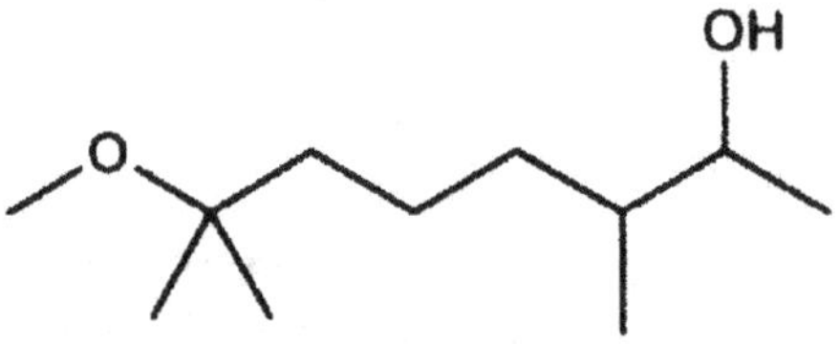

Osyrol

Frankincense (*Boswellia sp.*)

Frankincense has been used since ancient times as incense and as a special fragrance associated with divinity. It has also been reported to have very beneficial medicinal applications from antioxidant, antimicrobial, antibiofilm and anti-cancer biological activity. The essential oil is a very complex bouquet of 340 volatile chemicals and the composition is very variable depending on the species of *Boswellia* and the location and climate for propagation of the tree (9-13).

Hussain et al. (102) found that the oils most commonly contained terpenes such as α-and β-pinene, limonene, myrcene and linalool. However, these constituents were not always the predominate component of the oil, with some species exhibiting high concentrations of α-thujene, octyl acetate, octanol acetate and E-β-ocimene. The triterpenoid Boswellic acids with purported medicinal properties are too high of molecular weight to be part of the volatile components.

The aroma of frankincense is rather variable dependent on the factors that affect the chemical composition of the essential oil, but generally the odor is balsamic, sweet woody, green spicy with notes of lemon. The structures of β-pinene, linalool and an ocimene have already been shown. Myrcene, α-thujene and the complicated structure of boswellic acid are shown below.

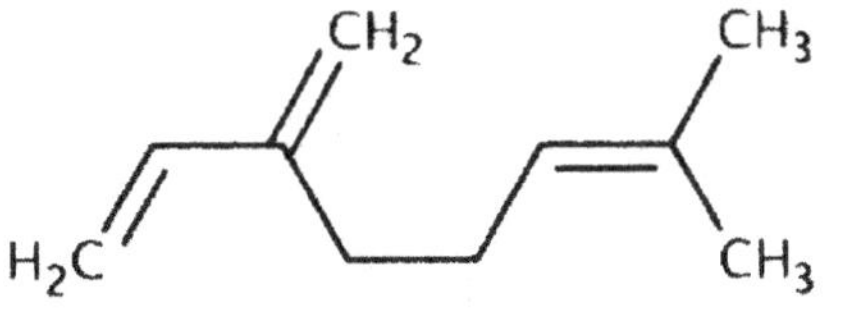

Myrcene

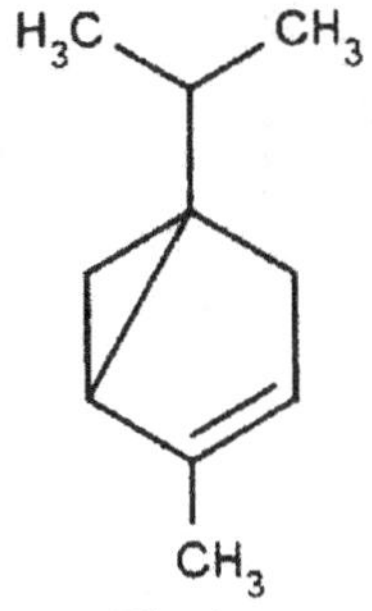

α-Thujene

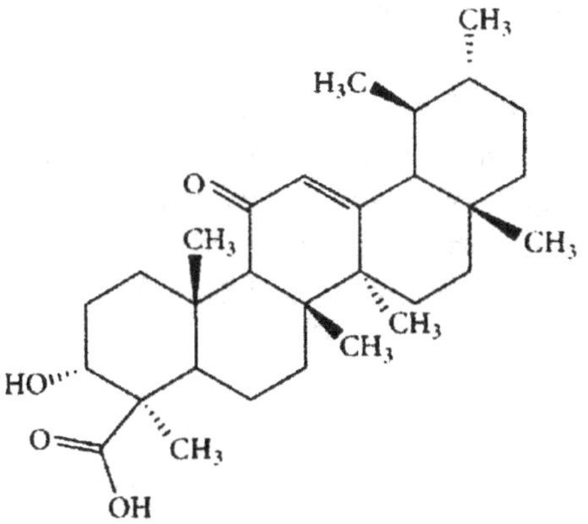

Boswellic acid

Labdanum (*Cistus ladaniferus*)

The resinous gum extract of labdanum has a tenacious woody-amber, leather-like odor and the oil is a composed of about 300 compounds with 186 identified so far. The odor is from a complex mixture of oxygenated constituents in rather low concentrations. The amber character is from Ambrox (0.7%) shown below and several other trace constituents, while the smoky-leather and hay-like notes are from a host of different phenols at 1.5% concentration. The moss undertones are from the same compound the gives oakmoss its aroma, Evernyl, shown below for oakmoss. One of the compounds responsible for the herbaceous note of labdanum is Tagetone also shown below (9-13).

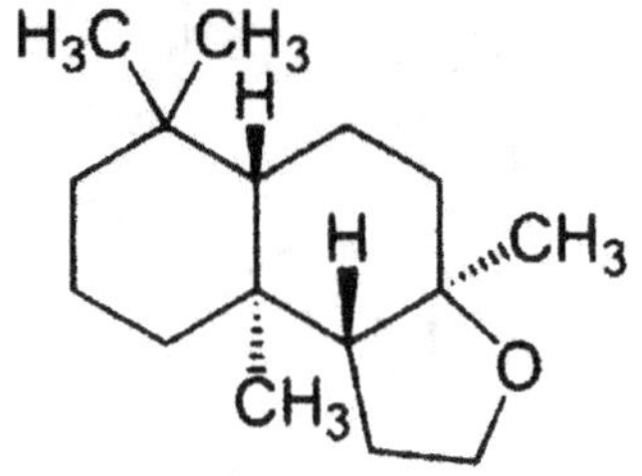

Ambrox

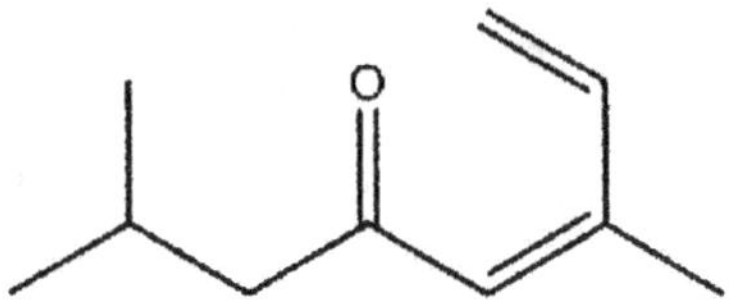

Tagetone

Galbanum (*Ferula galbanifkua*)

Galbanum is used as both a base note and often as a top "green" note. It is obtained by steam distillation of the air dried gum resin exudate from the tree. The oil is complex but just a few compounds in low concentration shown below, impart the characteristic odor. Galbanolene has a unique, transparent, greenish marine/metallic odor and Galbanum Pyrazine has a strong, earthy bell pepper aroma (9-12).

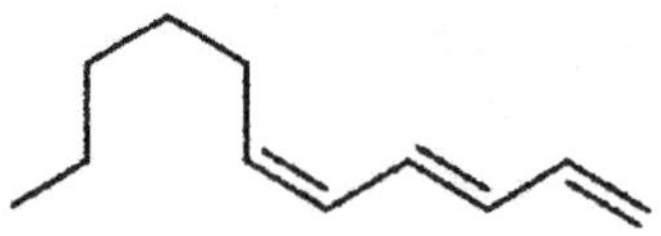

Galbanolene

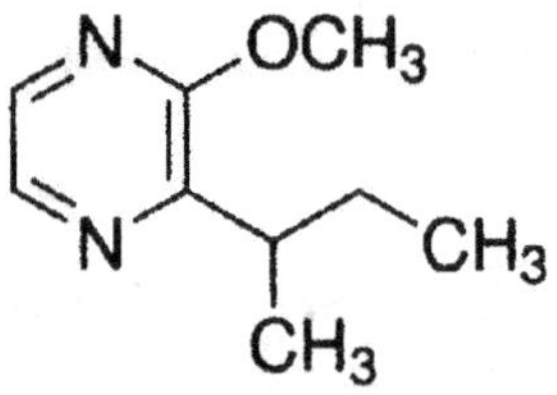

Galbanum Pyrazine

Several synthetic substitutes are now available in various molecular structures as shown below. Ally amyl glycolate has a strong, fruity galbanum scent with pineapple overtones and adds sparkle to *Drakkar Noir (Guy Laroche), Cool Water (Davidoff)* and *Egoïste Platinum (Chanel)* fragrances. Galbazine (Givaudan-Roure) has a green pepper aroma and Cyclogalbanate (Dragoco) provides a strong, fruity, herbal-green odor (12-15).

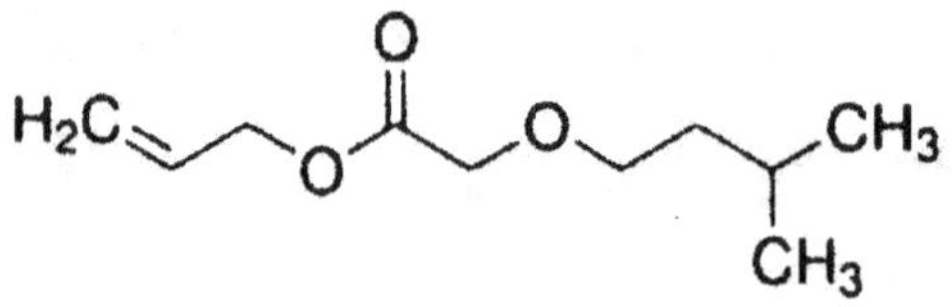

Ally amyl glycolate

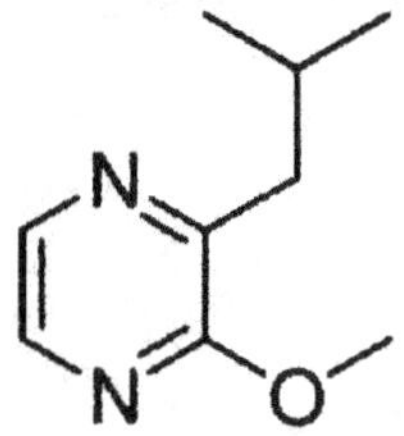

Galbazine

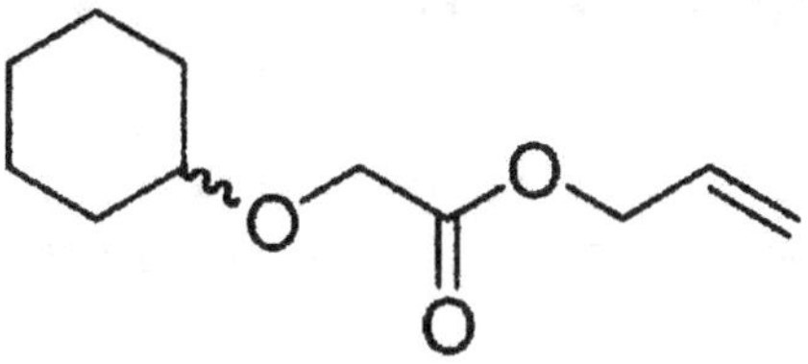

Cyclogalbanate

Oakmoss (*Everenia prunastri*)

Oakmoss is a primary component of Chypre and Fougère accord perfumes. The odor of oakmoss is mossy-woody with characteristic phenolic and smoky, almost animal-like nuances (8-10). The extracts from the lichen plant contain over 80 components and the constituents most responsible for the odor are Orcin derivatives and Evernyl (or Veramoss) shown below. However, as previously pointed out, oakmoss has been severely restricted for use in perfumery, to 0.1%, due to two skin irritating components of atranol and chloroatranol (shown below). Removal of these constituents from the extracts adds significantly to the cost such that there has been a significant decline in consumption of oakmoss (13).

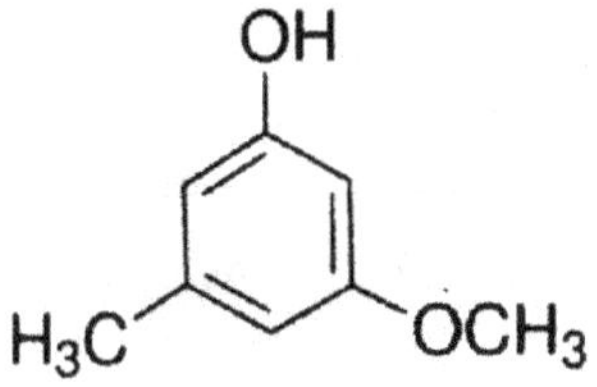

Orcinyl-3

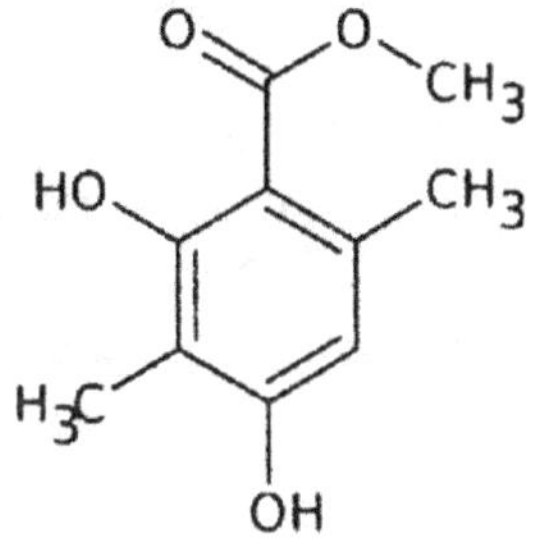

Evernyl

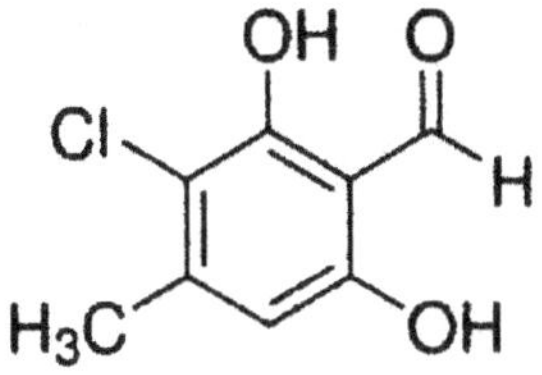

Chloroatranol

Synthetic Animal Odorants

Due to both ethical and legal restrictions odorants from animal sources are essentially no longer used in perfumery. As discussed in a previous section, the musk odorant was primarily obtained from the musk dear and civet from the civet cat, both of which have now been replaced by a variety of synthetics. Ambergris, from chunks of regurgitation material of sperm whales, is still found occasionally but it very rare and expensive and synthetic substitutes have also been produced for use in fragrances. A nature identical synthetic has not yet been produced for the castoreum, the odor extract from beavers (13,14,19,36,103-108).

Musk and Civet

Musks notes are very important to perfumery as a base note and are an essential component of practically all commercial perfumes. There are essentially no fragrances on the market that do not contain musk odorants. Musks harmonize the perfume, smoothing the rough edges, providing a lively, warm, soft almost powdery character to the fragrance. They are also an important fixative and substantially increase the longevity of the perfume (10,19,103).

The first synthetic musk was discovered in the late 19th century when the chemist Albert Bauer noticed a warm, sweet aroma when attempting to produce an analog for the explosive TNT (TriNitroToluene). Labelled the Bauer musk it led to synthesis of a whole series of related molecules termed the **Nitromusks**, a major class of musk compounds important to early perfumery shown below (13,14,26,105-107).

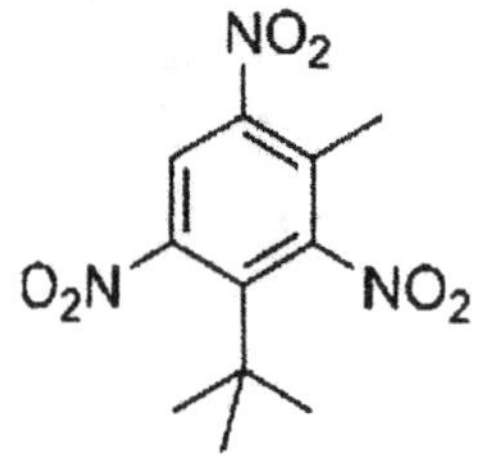

Bauer Musk

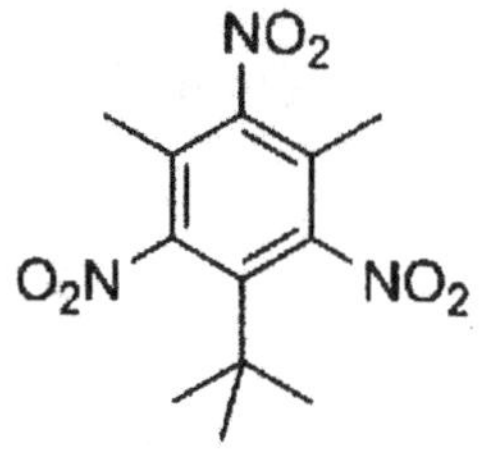

Musk Xylene

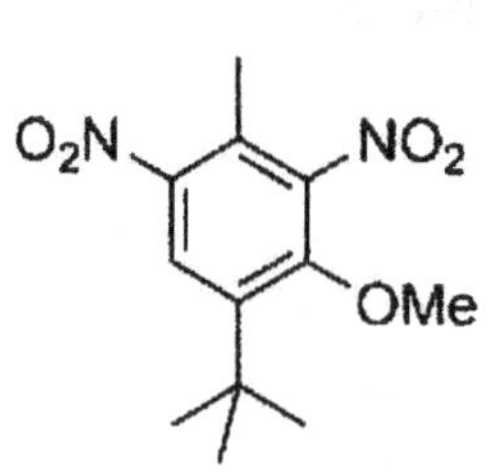

Musk Ambrette

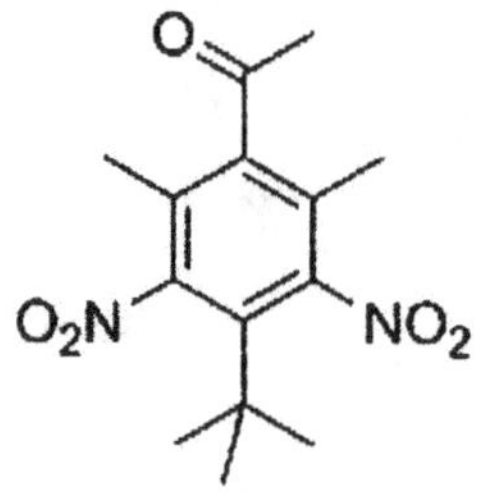

Musk Ketone

These musk compounds were used extensively for perfumery up to the 1960's. The odor of musk xylene is not as smooth as the others and was used more in soaps and detergents. The aroma of musk ketone best resembles natural musk, while musk ambrette has nuances of the floral, freesia. Musk ambrette and musk ketone were used to enhance the powdery-floral aspects in the original formulation of *Chanel No. 5,* in a combined concentration of 6% (13,19). Musk ambrette was also used with spicy eugenol for creation of the exuberant, *L'Air du Temps (Nina Ricci).*

The era of the nitro musks for perfumery came to an end when they were found to have adverse neuro- and phototoxicity activity. Their high reactivity also caused color changes in the perfumes over time. The search was on for new generations of synthetic musk molecules and two additional classes of musks, the **Macrocyclic Musks** and the **Polycyclic Musks**, were created by perfume chemists.

Macrocyclic Musks

The structure of the primary odor compound in natural deer musk, Muscone, was elucidated in 1939 by Leopold Ruzicka as shown below. It is obvious that it is a large cyclic molecule, macrocyclic. Similar musky compounds were found in civet extracts and in odorants of some plants as well. Civetone and Exaltone were isolated from civet while macrocyclic lactones, with a musky odor, were found in a number of plant species. Both *Jicky* and *Shalimar* fragrances by Guerlain have a prominent civet type note (13,14,26,105-107).

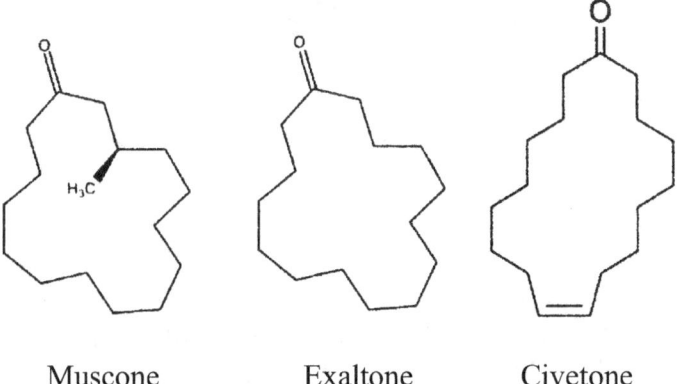

Muscone   Exaltone   Civetone

Macrocyclic musk lactones have an oxygen atom directly bonded in the ring structure and occur as Exaltolide in angelica root oil, oriental tobacco and several species of the Orchidaceae family. Ambrette seed oil also contains a macrocylic lactone, Ambrettolide. Although it is too expensive to isolate the macrocyclic compounds from these sources for perfumery, several were used as the basis for approaches to synthesize less expensive nature-like macrocyclic musks.

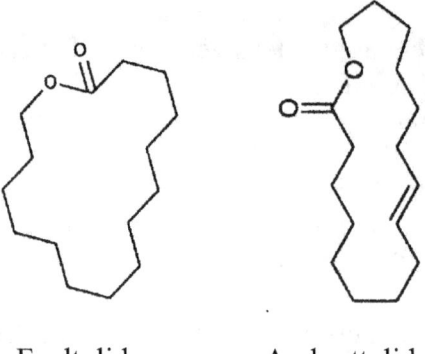

Exaltolide   Ambrettolide

Exaltone and Exaltolide were eventually synthesized at a dramatically lower price and marketed to the perfumery industry. Additional synthetic variants of the macrocyclic molecules were created as Muscenone (with powdery scent like that of the musk ketone), Nirvanolide and Cosmone, and the less desirable but also less expensive, Ethylene brassylate. The latter compound has woody, floral nuances and is often utilized in cosmetic products. Further tweaking of the position of the double bond in the structure of the macrocyclic compounds resulted in several more, new musk molecules such as Ambretone (Velvione) which has a soft velvet-like aroma similar to the older nitromusks (13,14,26,105-107).

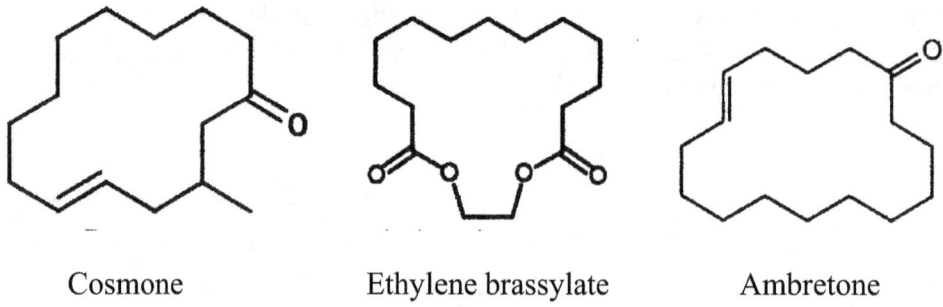

Cosmone   Ethylene brassylate   Ambretone

Polycyclic Musks

The first synthetic polycyclic musk (PCM) compound was marketed as Phantolide in 1951. This was followed by a series of new introductions of PCMs as Tonalide, Versalide, Vulcanolide, Celestolide and Galaxolide. These musks in general are relatively inexpensive and very persistent and as a result have been extensively incorporated into many contemporary fragrances (13,14,26,105-107).

As previously discussed, Sophia Grojsman created her new genre of "Horizontal Perfumes" using high concentrations of Galaxolide (21%) along with Iso E Super, Hedione (methyl dihydrojasmonate) and Ionones. The structure of Iso E Super is shown below and the structures for Hedione (see Aldehydes) and Ionones (see Violet) have already been shown.

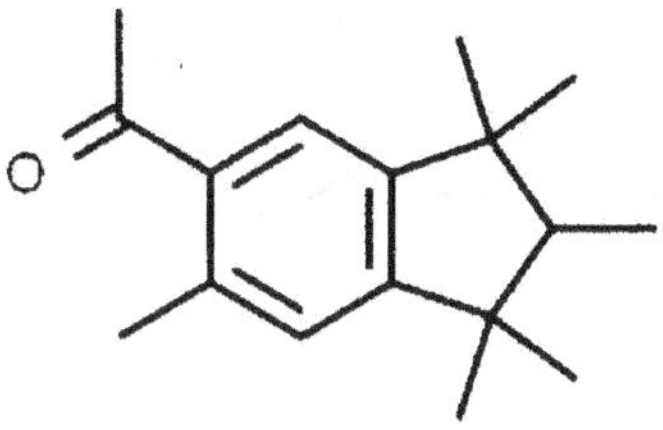

Phantolide

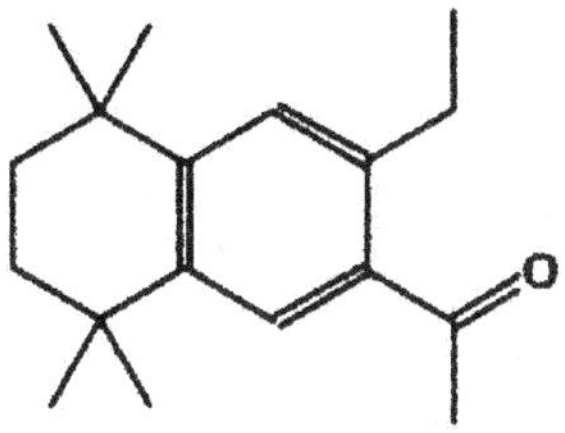

Versalide

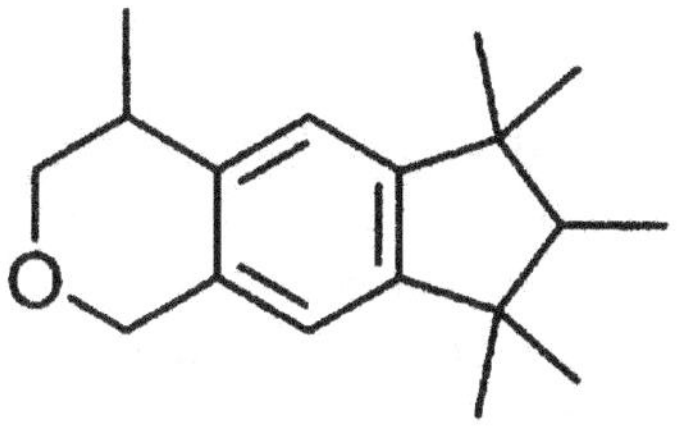

Galaxolide

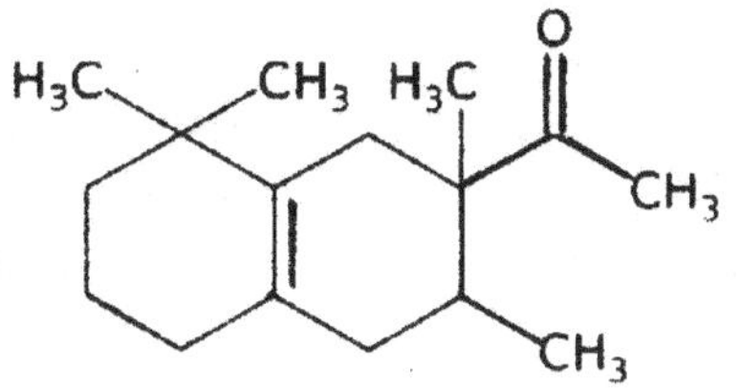

Iso E Super

There is another sub-class of PCMs that have been created which also have a polycyclic structure, but do not contain an aromatic benzene ring in their core. These molecules include Cashmeran, Moxalone and Nebulone. Cashmeran is the compound responsible for the "Cashmeran Wood" note in perfumes and has woody, ambery and leathery aspects in the musky aroma.

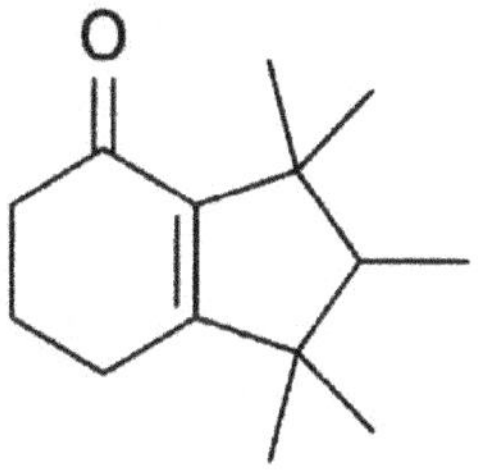

Cashmeran

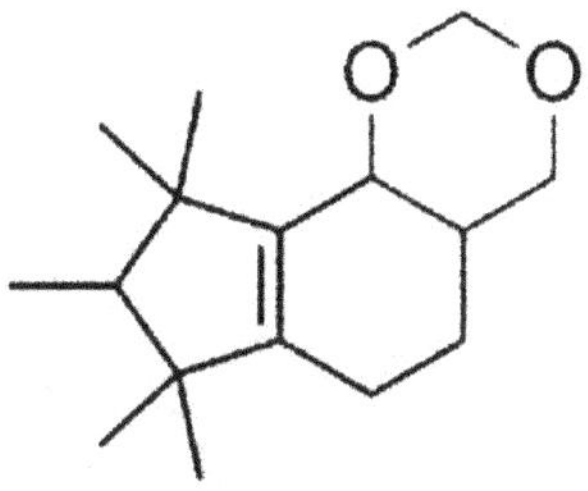

Nebulone

Two notable fragrances with the synthetic PCMs are *Lovely (Sarah Jessica Parker)* with 2.4% Nebulone and *Flower (Kenzo)* with 3.8% Helvetolide (13,17).

Linear Musks or Acyclic Musks

These are a newer generation of compounds for fragrance applications. Although the first linear musk, Cyclomusk, was introduced in 1975, it took another 15 years until the next musk was produced as Helvetolide. The chemists apparently creatively imagined a linear structure for this molecule that mimicked the circumference of the macrocyclic ring structures, resulting in the musky, floral and even fruity aroma of the Helvetolide molecule (13,14,26,105-107).

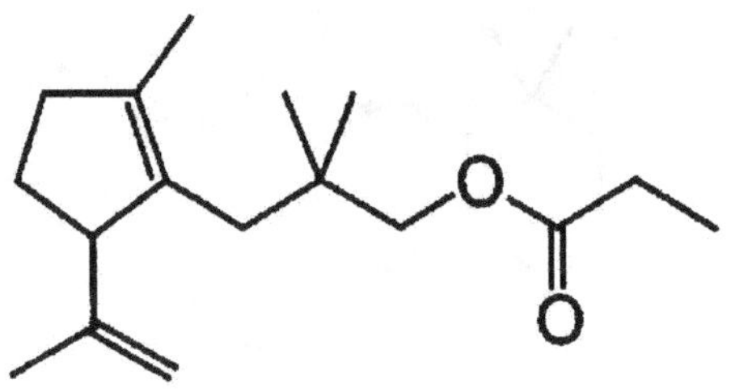

Cyclomusk 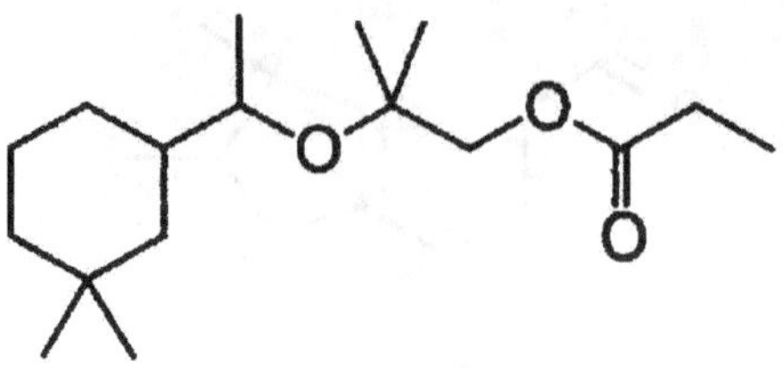   Helvetolide

Improved musk odors were realized by stabilizing the structure of the acyclic molecules by addition of carbonyl and ester functions to give Romandolide and Serenolide. This later musk, Serenolide, has five times the intensity of Helvetolide. The linear musks have been utilized to create the newer "White Musk" accords by incorporating Helvetolide or Serenolide with the macrocyclic musk, Habanolide, providing the soft, cotton-linen effects of the accord.

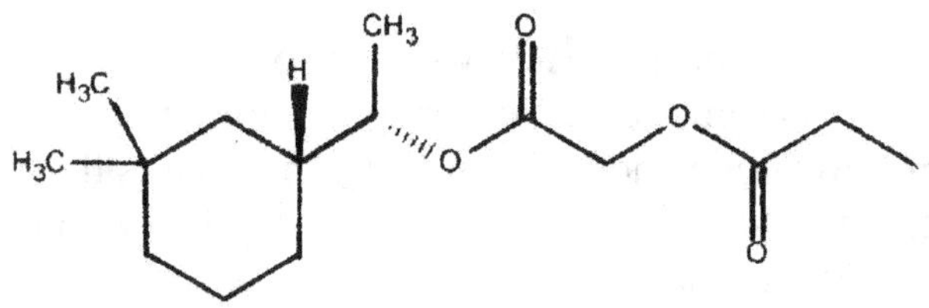

Romandolide

Dienone Musks

These are the most recent new musk odorants. The array of these acyclic musks have the added aroma of, for example, dried fruits, florals of violet and lily-of-the-valley and even beetroot. Further development of musk odorants are actively being pursued by the major perfume houses.

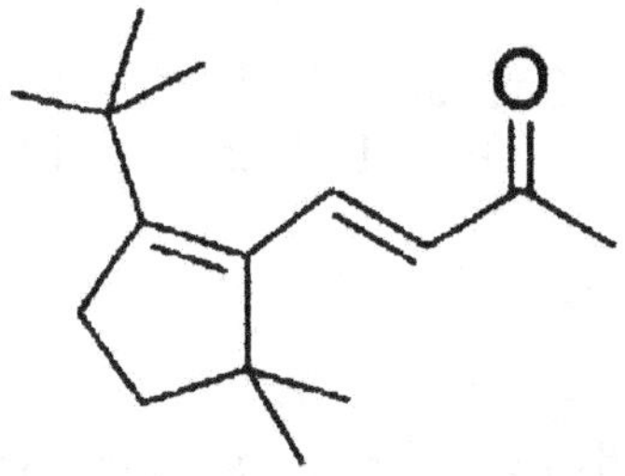

Dienone Musk

Ambergris

Collection of aged ambergris is very rare and most of the odorous molecules occur only in trace amounts in the extract (~0.3%) such that a variety of synthetic compounds are now routine substitutes. Two important odor compounds identified in ambergris were Ambrox and α-Ambrinol both of which have been produced by synthetic methods for use in perfumery. However, perfume chemists continued their work and produced several more ambergris odor analogs such as Amberketal used in the formulation of *Anaïs Anaïs (Cacharel)* and Spirambrene in *Kenzo pour Homme (Kenzo)*. The structure of these molecules is shown below (13,14,26,105-107).

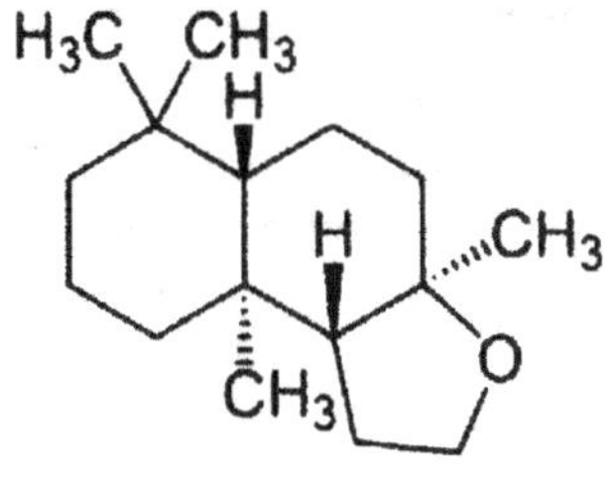

Ambrox 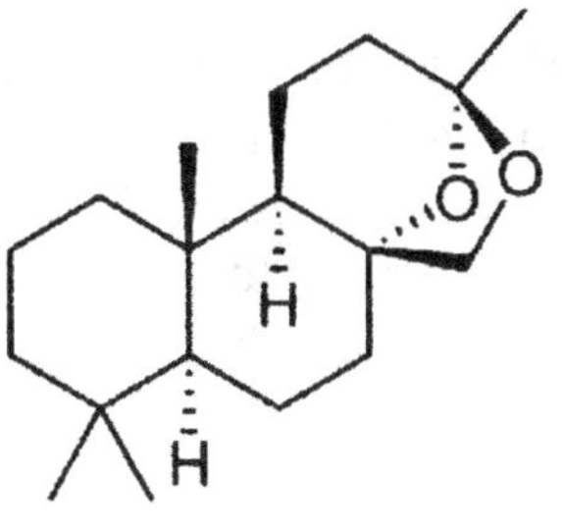   Amberketal 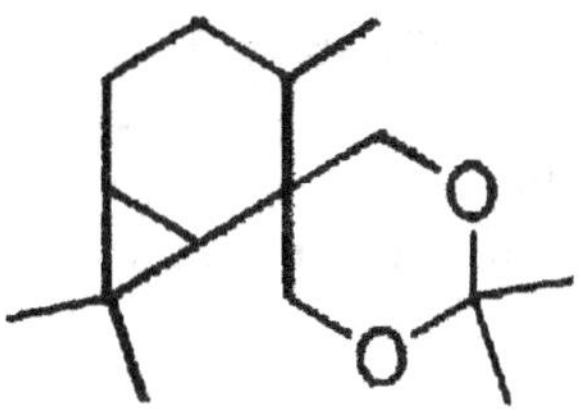   Spirambrene

Two more additions to the available synthetic ambergris notes are Suprambrox with a powerful, rich aroma and found in *Escape for Men* by Calvin Klein and the pared down structure of Ambermax that evokes a woody nuance in the amber note and used in the fragrance *Boss (Hugo Boss)*.

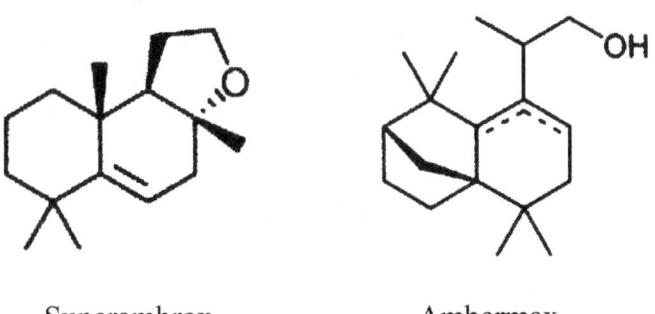

Superambrox   Ambermax

Chapter 5. Physiology and Theories of Smell

The human senses of touch, sight and hearing have been extensively studied and well understood, however the factors affecting the sense of smell of volatile chemicals is still under some debate. Several thousand odors can be distinguished by the sense of smell but it is difficult to categorize them based on chemical structure. In this section the proposed theories of smell will be presented and the most probable mechanism suggested.

Physiology of Olfaction

Odor quality is strongly influenced by memory, learning and culture. Olfaction or the sense of smell, is very sensitive and molecules can be detected at very low thresholds compared to taste. However, the odor receptors become rapidly desensitized, within one minute, to repeated exposure. Thus, our sense of smell is constantly ready for detection of novel odor molecules, but once recognized and adapted, the odor is essentially no longer distinguished (48,110). This information is known by professional perfumers and is important for both students of olfaction and fragrance consumers.

For odor detection, there are three critical characteristics that molecules must possess. They must be:
Volatile (low vapor pressure)
Low molecular weight (< 300)
Lipophilic (non-polar, soluble in fats/oils)

The volatile odorant chemical compounds are carried by inhaled air to the olfactory region located in the roof of the nasal cavities of the human nose, just below and between the eyes as shown in Figure 5.1. This area is only about 2.5 cm^2. The olfactory region consists of 8-20 cilia of 30-200 microns in length projecting down out of the olfactory epithelium into a layer of mucous only 60μ thick as shown in Figure 5.2. The mucous layer is lipid-rich (fatty, wax) and assists in transporting the odorant molecules to the receptors. The surface of the many cilia is the site where molecular reception with the odorant occurs and sensory transmission starts for olfactory reception (34,109,110).

The odor receptor proteins and their genes were discovered by Linda Buck and Richard Axel at Columbia & Harvard Universities for which they received the Nobel Prize in Physiology in 2004. It has subsequently been determined that there are 347 human receptors with a 7-fold helical G-protein structure (Figure 5.3). In comparison mice have 1,000 of these receptors (111-114).

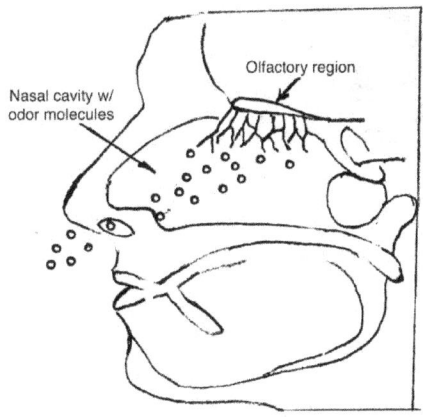

Figure 5.1 Human olfactory region.

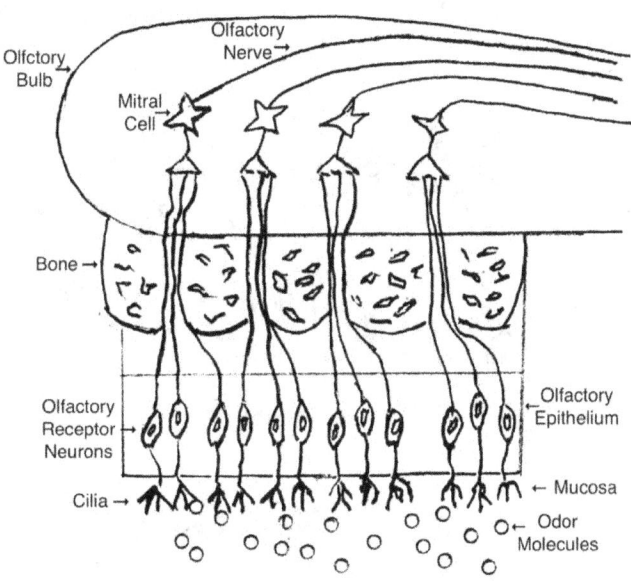

Figure 5.2 Close-up of olfactory region showing projecting cilia which interact with odor molecules for transmission to odor receptors (34,109,110).

In subsequent research, Zozulya et al. (115) reported the identification and physical cloning of the 347 putative human full-length odorant receptor genes. Comparative sequence analysis of the predicted gene products allowed them to identify and define a number of consensus sequence motifs and structural features of the family of receptors. They believe these sequences essentially represent the complete repertoire of functional human odorant receptors. This is an important step in understanding receptor-ligand specificity and combinatorial encoding of odorant stimuli in human olfaction.

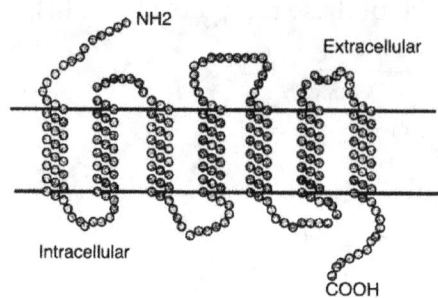

Figure 5.3 Seven-fold helical G-protein structure (112).

Figure 5.4 shows a cross-section of the mucous biological membrane containing the olfactory 7-helical G-protein receptor traversing the membrane. The cellular membrane is composed of phospholipids that impart a central hydrophobic core in a lipid bilayer of the long hydrocarbon chains, terminated with phosphate groups (shown as circular structures). The membrane is impermeable to large molecules and highly charged ionic molecules, but quite permeable to lipid soluble, low molecular weight molecules.

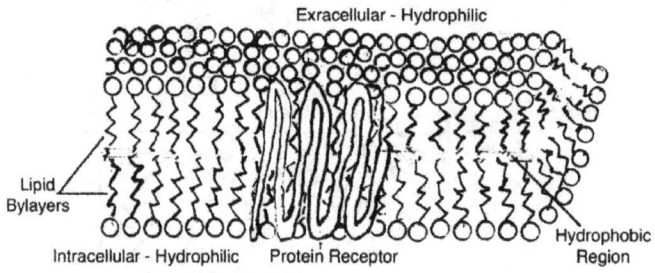

Figure 5.4 Cross-section of cellular mucous membrane containing receptor G-protein (34,109,110).

Search for New Odorants

The search for new odorants is motivated by several factors, one is commercial gain since the fragrance industry is a billion-dollar business and marketing requires both more stable yet biodegradable ingredients. There is also a need for new, unique odorants for development of modern trendy perfumes.

Another motivation is for health and environmental considerations since more and more odorants are banned each year due to environmental regulations. Since some people have adverse allergic and neurologic reactions to some chemicals in perfumes it is necessary to synthesize new molecules with the same smell to replace the ones in both classical and contemporary popular perfumes. However, it should be noted that it can cost up to $1 million from concept to production for a new material.

Theories of Smell

There is a pervasive need in the perfume industry for the ability to predict the odor of molecules from their structural characteristics. This would allow the chemist to construct new odorous molecules, not only based on known molecular structures, but provide the ability to conceive of new odor molecular materials.

In the 1946, Nobel Prize winning chemist, Linus Pauling, suggested that it was probable that the molecular shape of the volatile organic chemicals was the main factor affecting smell. This theory was held by a number of other scientists and extensive research was carried out for proof of the hypothesis. However, a second theory of smell was subsequently proposed based on the vibrational character of molecules. The evidence provided for all the current theories of smell are described in this section.

Molecular Shape Theory of Smell

The British chemist Amoore (116-118) was one of the early investigators on how molecular shape affects smell. He grouped the odorants into seven classes according to descriptions in the literature as camphoraceous, musky, floral, minty, ethereal, pungent and rancid. Amore proposed a type of "Lock and Key" model, similar to explanations for the action of enzymes for predicting odor as shown in Figure 5.5. He found similarity in the bulk three-dimensional structure of the molecules in each of the classes. Although his broad generalizations were not totally consistent with the variation in smell with molecular structure and could not be utilized for prediction of smell, his model for the camphoraceous odor has held the test of time with few exceptions.

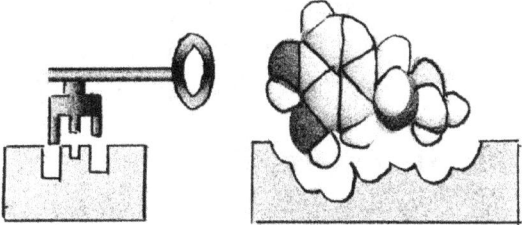

Figure 5.5 Lock & Key models for molecular recognition.

Several of Amore's "lock & key" structural models are shown in Figure 5.6. His rule for the camphoraceous odor specified that a hydrophobic molecule must be of ellipsoidal shape with a long axis of 0.95 nm and a short axis of 0.75, but the mechanism has not been deciphered. Research has continued on the molecular shape theory by a number of investigators since Amore's work in 1948 (119-121).

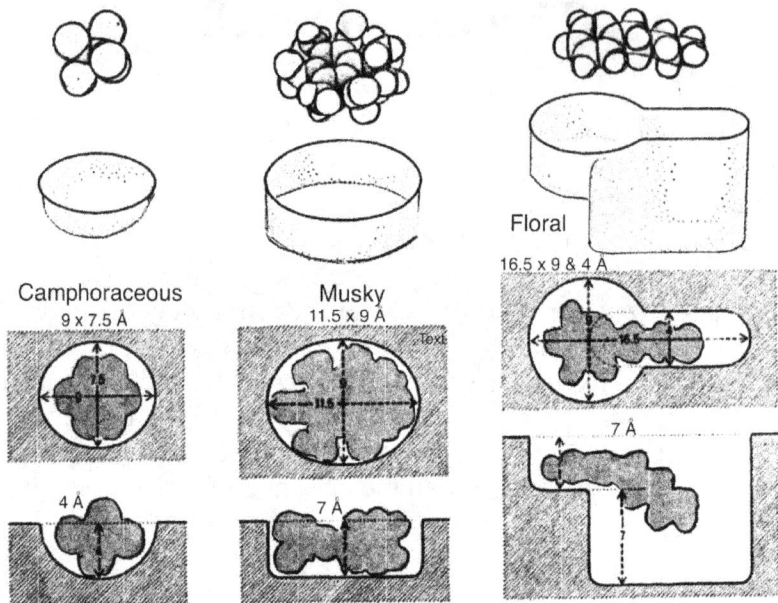

Figure 5.6 Amore's camphor molecular shape model (116-118).

It has also been observed that functional groups in the molecular structure can affect the odor. There are a variety of functional groups which are known to provide characteristic odors when part of molecular structure. For example, many alcohols (-OH), particularly terpene alcohols, have a fresh, strong, sweet floral aroma such as geraniol, linalool and phenylethylalcohol. The alcohol must have a least a C3 length of chain to impart the pleasant odors (13,14,26,105,122).

Many short chain carboxylic acid functional groups (-COOH) are known to impart pungent even rotten aromas as with formic and butyric acids. It is well known that the esters of the acids (-COOR) have the classic fruity aromas with banana for amyl acetate, pineapple for methyl butyrate and apricot for pentyl butyrate, to name just a few (see Figure 1.4 & Appendix I). However, as with the bulk dimensional approach, the use of just functional groups for odor prediction is not totally reliable.

Sandalwood Scent Structures

Sandalwood has received some of the most attention to crack the structure-odor relationships because of loss of the trees for production of the oil and the high cost and variable quality of the product. Hundreds of sandalwood molecules have been synthesized in the search for a cheaper synthetic alternative.

Considerable effort has been given to synthesizing a replacement for the molecule mainly responsible for the odor of sandalwood, (Z)-β-santalol. It can be done, but it takes eleven steps which is not commercially feasible. So continued research was directed towards altering structural parts of the molecule, both the ring structure and the branch chain (13-15,26).

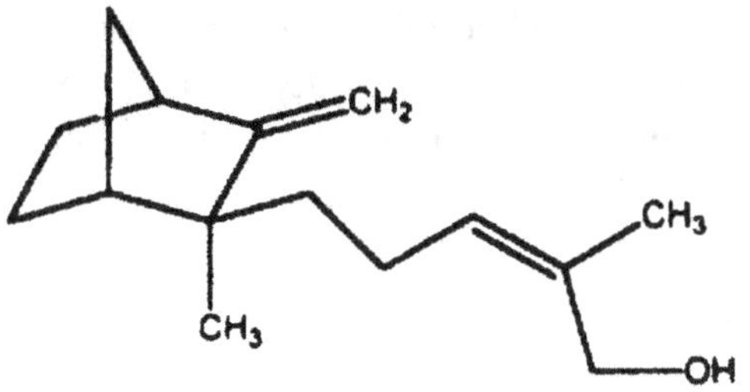

β-Santalol

Luca Turin (26) presented the workbook of the French chemist Jacques Vaillant from his efforts in the 1970's to produce a sandalwood scent. The structure of 45 newly synthesized molecules were on each page of the workbook with each molecule representing one-week of work. That amounts to about one year of work per page and only one molecule on this page smelled of sandalwood, with the others having odors of peach, cedarwood, lemongrass, rosewood, camphor and cut grass.

With continued hard work and a bit of serendipity a number of synthetic analogs of santalol are now available for use in perfumery. Santaliff by IFF was derived from a campholenic aldehyde and Javanol (Givaudan) was created by replacing two of the double bonds in the santalol structure with with cyclopropane ring structures. Also note the Osyrol structure, revealed in 1973, which does not contain a ring unit in the molecule and was a bit of a surprise to impart a sandalwood aroma (13,16,101,105).

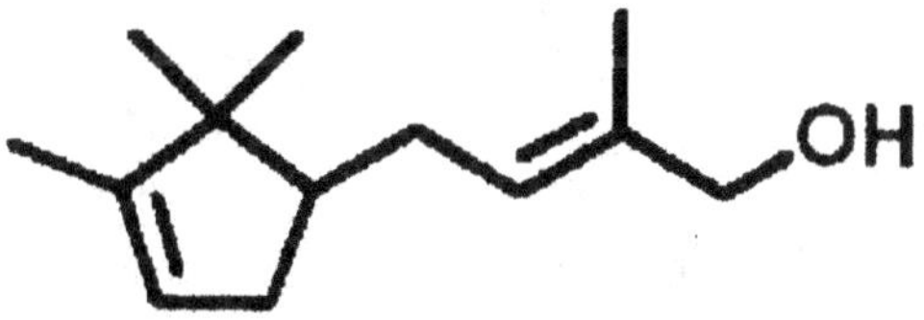

Santaliff 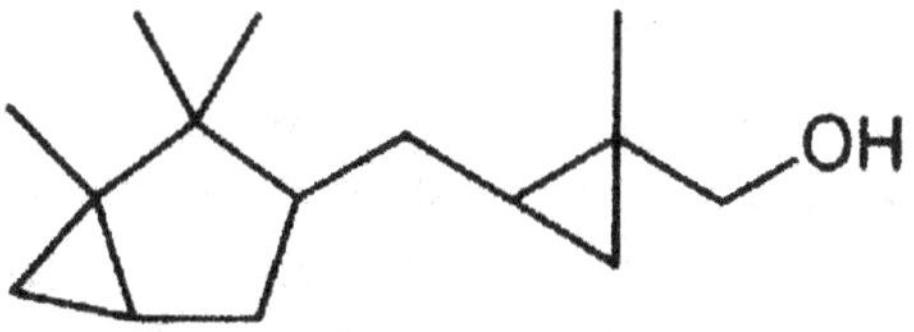   Javanol

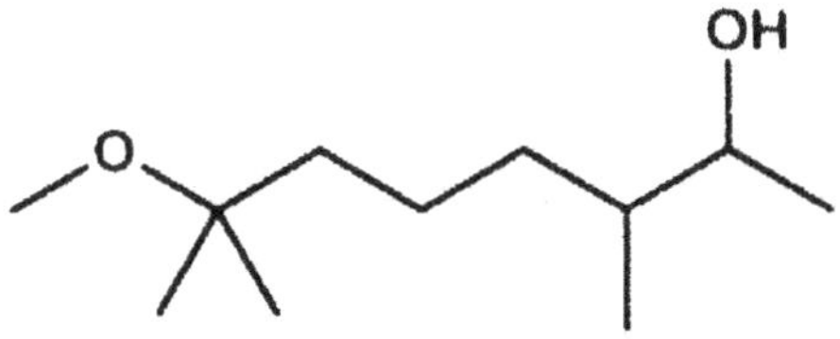

Osyrol

With all the various chemical structures synthesized to create a sandalwood scent, attempts were made by a number of investigators to develop predictive models for the sandalwood smell. The result has been sets of "odor rules" which can be more aptly referred to as guiding principles (13,105). The rules are based on molecular substituents and substitution patterns, intramolecular distances, molecular bulk, functional groups and relationships, polarity-dipole moments and molecular flexibility. Naipawer (123,124) proposed an odor rule for sandalwood in 1981 and several investigators have improved the predictability of the rule with adjustments to the parameter distances. The dimensions proposed by Buchbauer (125) and Dimoglo (126,127) for sandalwood odor are shown in Figure 5.7.

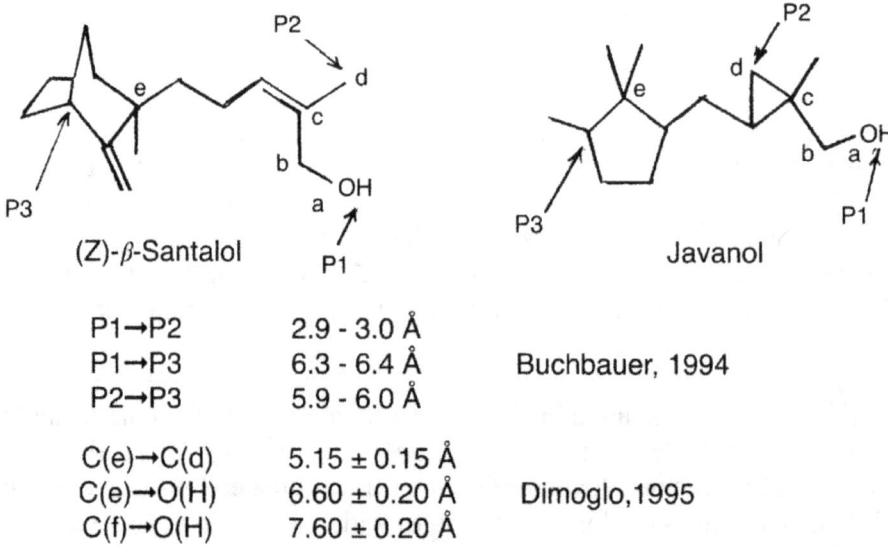

Figure 5.7 Sandalwood odor rule (13,125,127).

Additional odor rules have been devised for a number of other odorants including jasmine, rose and vetiver, but the most reliable are for sandalwood, ambergris and musks. However the utility of the rules is restricted to just the specified family of odorants and are not viable beyond the particular families. The empirical rules also have exceptions, for example, the molecule Osyrol which breaks the sandalwood rule even though it has a definite sandalwood scent.

Quantitative Structural-Activity Relationship (QSAR) Olfactory Models

In the past, research to discover new odorous molecular structures has been focused mainly on altering/modifying known molecules from nature along with an element of serendipity to realize new odorants. The development of the odor rules provided valuable direction for these investigations but they are only applicable within a "family" of odors as discussed above.

A more recent and sophisticated approach to molecular structure-property relationships for olfactory has been the use of Quantitative Structural-Activity

Relationship (QSAR) models. QSARs are predictive models derived from application of statistical tools correlating structural parameters with biological activity and have been useful in many areas of pharmacology. These models utilize the surface electronic properties of molecules and physiochemical parameters such as molecular volume, bulk of substituents, polarizability, etc. (13,14,56,105,128-132).

The olfactophore QSAR models are produced with similar software packages as those utilized in pharmacology. The supposition is that molecules with similar shapes, charge distributions and polarizability will be recognized by the same proteins. They are conceptual design tools that work best when the structure of both the molecule and the receptor are accurately known. So far for perfumery, they have been based on the structure of the odorant alone, but with advances in the structure and action of the G-protein-coupled receptors the accuracy will be greatly improved by additional application of models of the receptor binding cavity. Thus, the chemical structure can be correlated with the affinity towards the receptor. However the task is formidable since matches would be required for all the roughly 347 human olfactory receptors (13,14,5,115).

However, the olfactophore models allow a much better three-dimensional insight into the structural characteristics required to reproduce an odor as compared to the Odor Rules. As an example, by combining all the previous information available for sandalwood, an olfactophore model structure was produced and a sketch of this model is shown in Figure 5.8 (13,14,56,128). Shown is three-dimensional overlap of (Z)-β-santalol from sandalwood oil with the synthetic, Javanol, which has a strong sandalwood odor (13).

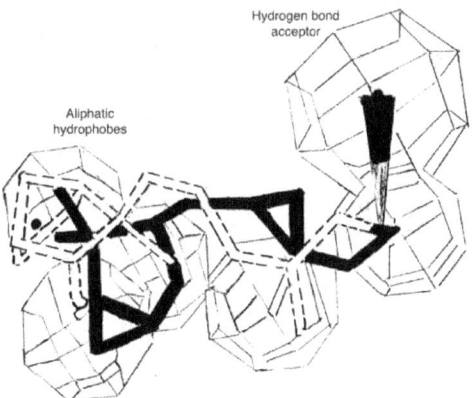

Figure 5.8 3-D overlap of (Z)-β-santalol (dotted line, no fill) with synthetic Javanol (black, solid fill) (modified from Ohloff et al.,13).

Most structure-odor relationships have been developed in hindsight and there are exceptions to most of the rules. Even these QSAR models are not useful outside the scope of the data set so they are still not useful for the discovery of new classes of compounds. However molecular modeling has provided superior guidance for the fragrance chemist and has proven useful for production of fragrance ingredients such as Javanol, Rossitol, Belambre and Azurone and helpful for development of Marine and new Lily-of-Valley (Muget) fragrances (14,56).

Problems with Molecular Shape Theory

As we have seen, the empirical odor rules are full of exceptions and open to challenge. The use of QSARs is also helpful but not sufficient for discovery of new classes of scents across odorant families. Another problem with the molecular shape theory is that some odorants smell the same even though they are structurally different. Figure 5.9 shows a variety of different compounds that all smell of musk with the macro-musks distinctly different in structure compared to the nitro-musks. Similarly, Figure 5.10 shows several molecules of different structures that all have a muget odor, particularly noticeable in this example is the molecule at the bottom with the ring structure containing a sulfur atom (13,14,26,56,105,131,132).

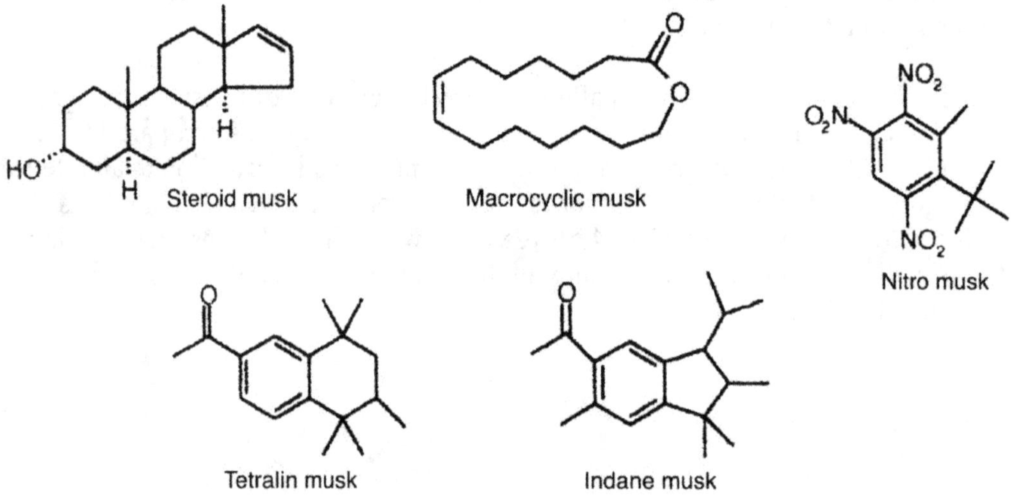

Figure 5.9 Musk molecules (13-15,26).

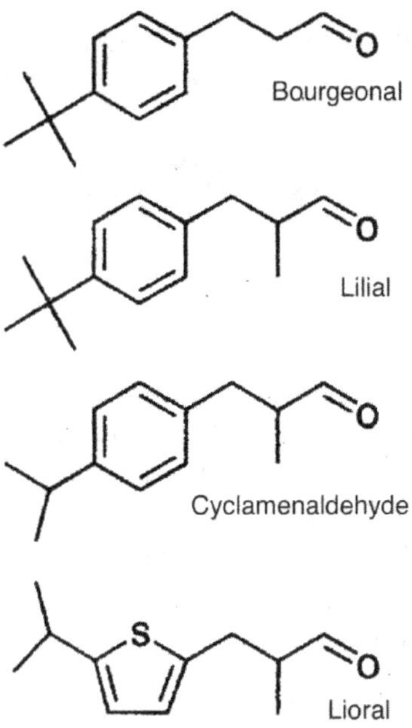

Figure 5.10 Muget molecules (13-15,26).

Another striking example of different molecular structures having the same odor is for benzaldehyde and hydrogen cyanide, both of which smell of almond.

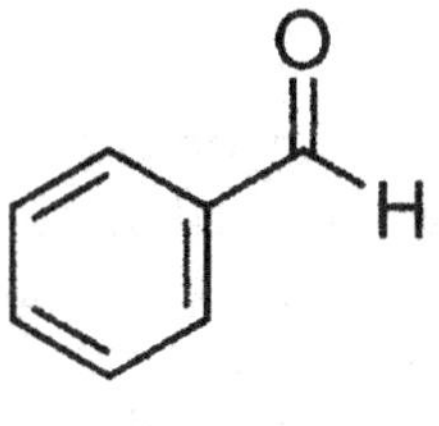

Benzaldehyde   Hydrogen cyanide

Turin (26) has also pointed out the reverse phenomenon, based on space filling models, that some molecules with the same shape can smell differently. He compared Pinanethiol and Pinanol which have very similar 3-D shapes but have different odors. The pinanethiol contains a sulfur moiety while the pinanol is an alcohol which can dramatically affect the odor. Turin presented another example for the same shape but different odors for Ferrocene and Nickelocene (26).

In summary, there are clearly some problems associated with the molecular shape theory for odor as follows:
- Amore's lock and key model does not work for many odor descriptors
- Molecules with dramatically different structure can smell very similar and
- Odor rules have many exceptions and QSAR analyzes also only apply to specific

families of odors.
As a result, an alternate theory of smell, the Vibrational Theory, was proposed many years ago and recently promoted by Turin (26,106-107,133-136).

Vibrational Theory of Smell

One of the earliest proponents of the vibrational theory of smell was Dyson in 1938 (132). The concept is that odor is created by the natural vibration of polyatomic molecules which show many modes of movement between atoms, including asymmetric and symmetric wagging, twisting, scissoring and rocking. Evidence for the vibrational theory was provided by several additional investigators (119-121,133,134) but the strongest premise has been provided by Turin based on the concept of Electron Tunneling (26,106,107,135,136). Turin proposed that electron tunneling was a possible way for receptor identification of odorants in biological systems. The discovery and mechanism of electron tunneling is described below.

Electron Tunneling

The phenomenon of electron tunneling was discovered in 1960's by Lambe and Jaklevic (137). They were studying tunneling effects in metal-oxide-metal junctions. Instead of finding band structure effects of the metal electrodes as they expected, they observed patterns related to vibrational impurities (organic material) contained in the insulating oxide layer.

They noticed mysterious kinks in the current, as a function of voltage, in the oxide layer between the two metal plates. As the voltage passed certain thresholds values, the conductance increased suddenly. When plotted in a step fashion they noticed the that pattern looked remarkably like an infrared spectrum of an organic compound. They were right, using inelastic electron tunneling they had measured the vibrational spectrum of organic compounds that had been trapped in the tunnel barrier during fabrication (137).

The electron tunneling event occurs when emitted phonons fills the gap. There are two modes of electron tunneling that occur when the electrons tunnel from one electrode to the other as shown in the Figure 5.11. In elastic electron tunneling the electrons tunnel from one electrode to the other without losing kinetic energy. They do not interact with the molecule in the system. Whereas, in inelastic electron tunneling the electrons donate energy to the contaminant molecule, exciting a vibrational mode and creating a new tunneling pathway. Thus, in the inelastic process some of the tunneling electrons lose energy by exciting vibrations of the adsorbate, which gives additional current contribution and leaves behind a vibrational quantum of energy (26,106,107,136,138).

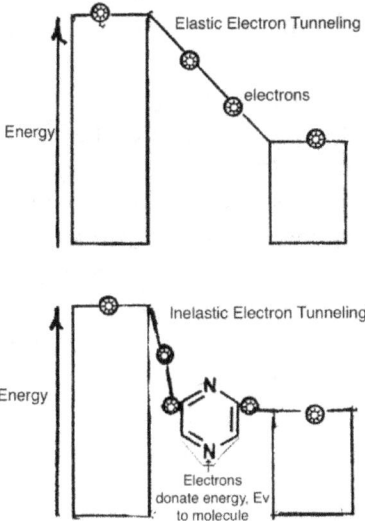

Figure 5.11 Electron tunneling (Turin, 26,106,107).

It is notable that inelastic electron tunneling spectroscopy has since been developed into powerful tools for surface characterization by Scanning Tunneling Microscopy (STM), chemical identification by Atomic Force Microscopy (AFM) and for atomic bonding investigations. The mechanism is also plausible for Biology.

Evidence for Vibrational Theory

Turin (26,106,107,136) has presented several categories of evidence for the vibrational theory of smell described in the following sections.

1. <u>Some odorants smell the same though structurally different</u>. It is known that compounds with sulfur-hydrogen bonds have a very noticeable "rotten eggs" odor such as methyl mercaptan and the thiols in skunk spray (2-buten-1-thiol and 2-quionolinemethanthiol). Turin noted that they all have an infrared (IR) vibration band at 2500 cm^{-1}. He then searched for other compounds which exhibited an IR band in this same region. Based on calculations using atomic masses and bond stiffness he determined that boranes should have a vibration in this same region and predicted that the odor should be the same. The problem is that boranes are some of the most volatile, highly flammable and toxic compounds in existence, however, he did discover one suitable for the smell test, decaborane, which had the same rotten egg smell.

2. <u>Some odorants smell different but have the same shape</u>. Turin noted that two compounds, Ferrocene and Nickelocene, have very similar three-dimensional shapes, so based on the molecular structure theory they should have similar odors. However, the two compounds smell differently and their strongest vibrational mode, involving an internal motion of the metal ion between the two rings, is also

different. Ferrocene has a spicy-camphoraceous odor and an IR band at 478 cm^{-1} while Nickelocene has an oily chemical smell and an IR band at 355 cm^{-1}.

3. <u>Deuteration does not affect shape but alters smell</u>. The hydrogen atoms of odorants can be exchanged with the isotope, deuterium or "heavy hydrogen", to give a deuterated molecule which has the same shape but with all the hydrogens replaced with deuterium. Deuteration does not alter atom size, bond length or stiffness. Thus according to the molecular shape theory, the odor should remain the same for both isotopes. To test the supposition, cyclopentadecanone (28H) was deuterated (C-D) and was found to have a nutty, roasted, burnt odor compared to the un-deuterated compound (C-H) which had a musky smell. This was attributed to the effect of the deuteration on the vibrational pattern with the C-D vibration now at 2200 cm^{-1} compared to the C-H vibration at 3000 cm^{-1}. However, Turin also noted that smaller molecules with eight carbons or less did not show a change in odor with deuteration.

4. <u>Fruit Flies (*Drosophila melanogaster*) ability to differentiate deuterated odorants</u>. Fruit flies possess a relatively well understood olfactory system, they exhibit keen olfactory discrimination and can be conditioned to selectively avoid or seek odors in established standard T-maze pattern methodology. This ability is also noted for other insects, such as cockroaches, and fish.

Franco et al (139) reported that fruit flies are attracted to acetophenone but are not attracted to the deuterated isotope and the aversion to the latter isotope increases with the number of deuterium atoms. The flies could also be trained with electric shocks to avoid the deuterated molecules or even to prefer them to the proton substituted molecules. These experiments indicate that fruit flies can smell deuterium no matter what the molecule. When the flies were exposed to nitriles which have a similar vibration to the C-D bond, the flies trained to avoid deuterium similarly avoided the nitrile, which supports the concept that the flies are smelling vibrations.

The evidence shows that fruit flies clearly differentiate between the isotopic odorants, that they can be conditioned to avoid the common or deuterated isotopes and can exhibit selective aversion to unrelated molecules with a vibrational mode in the C-D stretch region, favoring the vibrational theory of smell.

Dilemma of Molecular Mirror Images

Many molecules are chiral meaning they have mirror images, that is, they have left- and right-handed forms and are termed enantiomers. This distinction is very important in pharmacology and drug molecules are synthesized for a chosen handedness for proper efficacy.

There are very many enantiomeric odorants. If shape matters all pairs should smell different, but if vibration dominates, the pairs should smell the same. Of 400 odorants it was found that 60% smell alike and 40% smell different (26,106,107,136).

An example is the molecule carvone where the enantiomer R-Carvone has a spearmint odor and the S-Carvone has a caraway odor. Turin suggested that the mirror images for carvone are oriented in different directions in the receptor causing the different smells. He felt his explanation was supported by his addition of a carbonyl containing compound, pentanone, which transformed the smell of the spearmint enantiomer to a caraway odor. Brookes provided evidence that enantiomers only smell different when the molecules have some type of flexibility (140,141).

Problems with Vibrational Theory

It should be noted that there have been objections to the vibrational theory of smell by a number of scientists, some of whom were not able to reproduce all of the results from Turin's research (111-114). The vibrational theory clearly does not account for all the differences in smell and has not been used broadly for prediction of odor, however; it may have some role in the final elucidation of the quandary of odor perception.

Combinatorial Odotope Theory of Smell

Considerable progress has been made on the mechanisms of odor detection during the past 20 years, especially through the work of Linda Buck and Richard Axel who received the Nobel Prize for their contributions in this area in 2004 (112). They noted that the discriminatory power of the olfactory system, which governs the sense of smell, is immense since even slight changes in the structure of an odorant can alter the perceived odor quality.

Buck et al. (111-114) utilized mice for their experiments on olfactory perception. Their goal was to determine how information from different odorants are encoded in the nose. They exposed mouse olfactory receptors to a range of odorants and used calcium imaging to detect which receptors were stimulated by a particular odor. They utilized four different classes of aliphatic compounds with 4-8 carbon atoms and different functional groups for detection of neuron response. The classes of compounds were alcohols, carboxylic acids, bromocarboxylic acids and dicarboxylic acids. If a response was noted for a particular odorant it was further tested at a lower concentration.

Figure 5.12 shows the responses from 14 different neurons, and therefore the recognition profile of the odorant receptors obtained by Buck et al. (111-114). The size of the circles represents response intensity. The data reveal three important features of odorant reception. The first is that one receptor can recognize multiple odorants, the second is that one odorant can be recognized by multiple different receptors and the third and most important is that different odorants are recognized by different combinations of receptors. This indicated that the odorant reception family is used in a **Combinatorial Way** to encode the identity of different odorants.

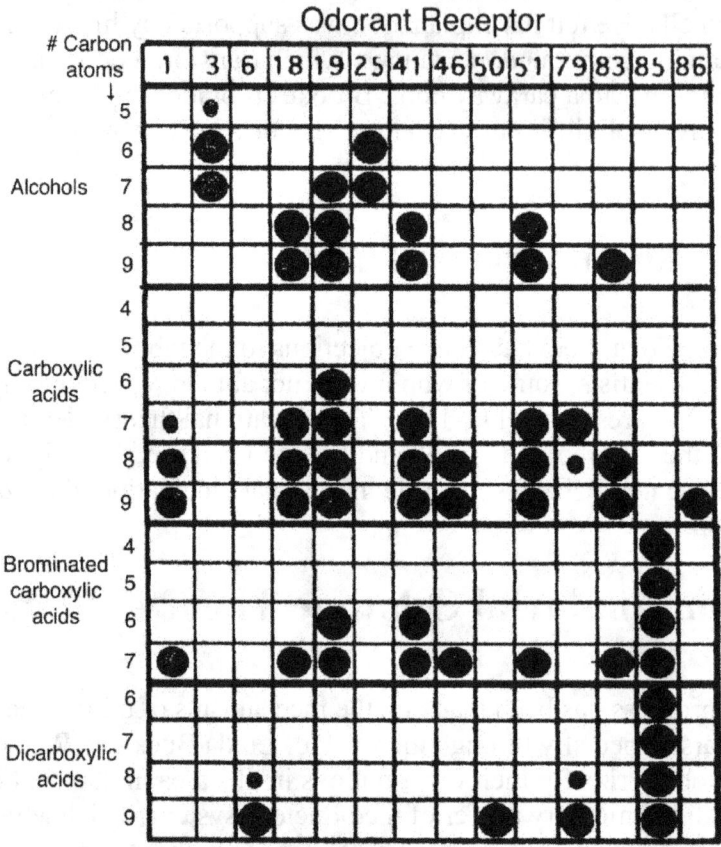

Figure 5.12 Responses of receptors to different odorants (Buck et al., 111-114).

Different odorants are recognized by different combinations of receptors but each receptor serves as one component of the codes for many different odorants. Different odorants have different receptor codes as shown in Figure 5.13 (14). Buck's work demonstrated that olfactory receptors are not tuned to specific odors and that the odorant signal to the brain constitutes a specific pattern of firing across a range of receptors in a combinatorial fashion. At higher concentrations, additional ORs were invariably recruited into odor response which explains why changing odor concentration can alter perceived odor.

This coding scheme allows for discrimination of an almost unlimited number of odorants based on the number of receptors. Only three ORs could generate almost a billion different codes. The scheme explains why changing the structure of an odorant even slightly can alter its perceived odor. The odor of alcohols in their sequence generally ranged from sweet, herbal, woody to fresh-floral while the acids ranged from rancid-sour to cheesy-nutty.

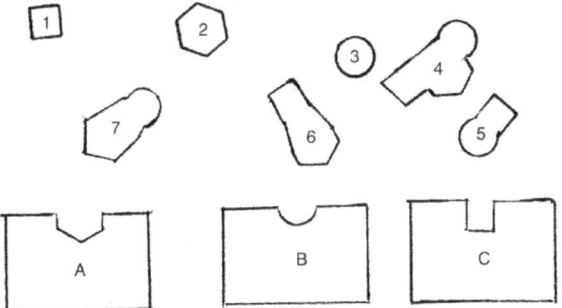

Figure 5.13 Models of odorants and receptors. Odorant 1 triggers receptor C; odorant 2 triggers receptor A; odorant 3 triggers receptor B; odorant 4 triggers all three receptors A, B & C; odorant 5 triggers receptors B & C, odorant 6 triggers receptors A & C, and odorant 7 triggers receptors A & B. Viewed in another way, receptor A triggers odorants 2, 4, 6 & 7, receptor B triggers odorants 3, 4 & 5 and receptor C triggers odorants 1, 4, 5 & 6 (modified from Sell, 14).

Olfactory Notes & Molecular Complexity

Research by Kermen et al. (147) is concordant with the combinatorial scheme of Buck et al (111-114). They found that the pleasantness of odorants and the number of olfactory notes are influenced by an odorants molecular complexity. Shown at the left in Figure 5.14 are several molecules of increasing molecular complexity measured by the diversity of atoms, symmetry, and bond connectivity (indicated in parentheses under the molecules). Shown at the right are the number of olfactory notes used to describe the odorant molecules. The odorants can be described by a few or many olfactory notes.

The results show that the more structurally complex an odorant, the more numerous the olfactory notes. It is also apparent from this investigation that complex odorants activate more olfactory receptors and hence evoke a greater number of olfactory notes. Another observation from this study was that odorants of low complexity, that evoke fewer notes, were perceived as more unpleasant and may serve as a warning cue from the olfactory system, such as low molecular weight carboxylic acids.

Compound (Molecular Complexity)	Number of Olfactory Notes
Furan (23)	3 - spicy, smokey, cinnamon
Allyl propionate (87)	4 - sour, fruity, apple, apricot
Eugenol (145)	4 - spicy, dry, peppery, woody
Coumarin (196)	6 - tobacco, nut, herbaceous, spicy, sweet, hay
Acetoeugenol (225)	7 - sweet, spicy, fruity, earthy, balsamic, leafy, floral

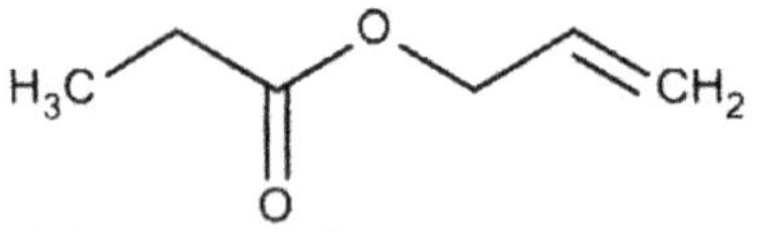

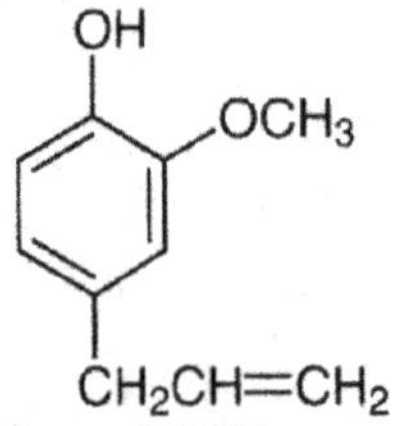

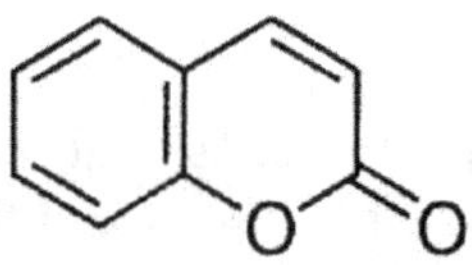

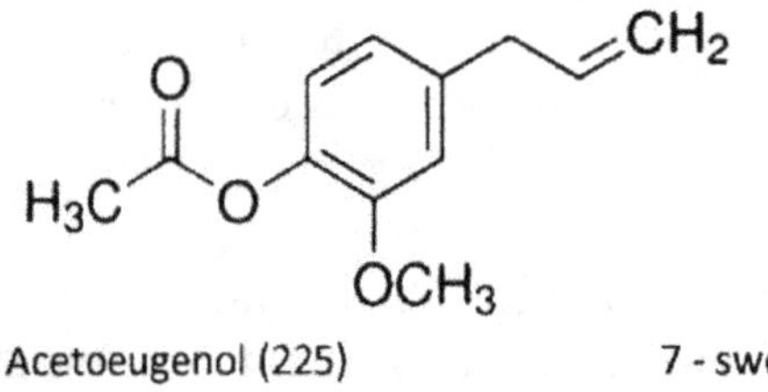

Figure 5.14 Effect of molecular complexity on number of olfactory notes (147).

Swipe-Card Theory of Smell

Jennifer Brookes of the University College London posed another provoking premise to the concept of odor perception with her Swipe-Card theory which provides a combination of both vibrational and molecular shape theories of smell. She used the analogy to key cards used for access to hotel rooms (Figure 5.15) (148,149). An approximate fit between an odorant and receptor (shape) is necessary to swipe key into a lock, but an internal message (vibration) is necessary to open door. Strong mixing of electronic and vibrational states is required with both states affected by molecular structure. For the swipe-card theory a compromise of geometrical (shape) factors is required in combination with the right energetic (vibrational) factors for receptor responses.

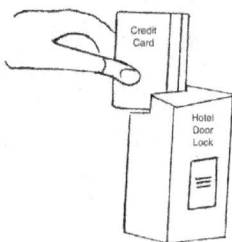

Figure 5.15 Swipe card room key.

The Mechanism of Smell is very complex and involves aspects of molecular structure with possible vibrational contributions. By far the best explanation of smell to date is the **Combinatorial Coding System** involving multiple receptors. Each type of receptor responds to a variety of odorant molecules and each odorant molecule interacts with a variety of receptor types.

References and Bibliography

1. E. Guenther, 1996, *The Essential Oils*, Krieger Pub., Melbourne, FL.
2. A. F. Hill, 1952, Economic Botany, McGraw-Hill, NY.
3. S. Arctander, 1994, *Perfume and Flavor Materials of Natural Origin*, Allured Pub., Carol Stream, IL.
4. K. Base & G. Buchbauer, 2010, *Handbook of Essential Oils: Science, Technology and Applications*, CRC Press, NY.
5. D. Williams, 1996, *The Chemistry of Essential Oils*, Micelle Press, Dorset, England.
6. P. Tongnuanchan & S. Benjakul, 2014, "Essential Oils: Extraction, Bioactivities and their Uses for Food Preservation," J. Food sci., 79 (7): R1231-R1237.
7. R. Guba, 2002, "The Modern Alchemy of Carbon Dioxide Extraction," Intern. J. Aromatherapy, 12 (3): 120-126.
8. S. Herman, 2017, "Fragrance" In: *Cosmetic Science and Technology: Theoretical Principals and Applications*, H. Maibach, et al., eds., Elsevier Pub., Philadelphia.
9. B. Jensen, 2020, Small Guide to Nature's Fragrances (bojensen.net)
10. J. Rose, 2020, Essential Oil Profiles (jeanne-blog.com)
11. Perfumes/Notes (fragrantica.com) & J. Moran, 2015, *Vintage Perfume*, Sunny Palms Press, Beverly Hills, CA.
12. The Good Scents Co. Information System, 2021 (thegoodscentscompany.com)
13. G. Ohloff, W. Pickenhagen & P. Kraft, 2012, *Scent and Chemistry: The Molecular World of Odors*, Wiley-VCH, Zurich.
14. C. S. Sell, ed., 2006, *The Chemistry of Fragrances*, 2nd ed., RSC Pub., Cambridge, UK.
15. R. Calkin & J. Jellinek, 1994, *Perfumery: Practice and Principles*, John Wiley & Sons Pub., NY.
16. R. Dove, 2014, *The Essence of Perfume*, Black Dog Pub., London.
17. N. Chapman, 2019, *Perfume: In Search of Your Secret Scent*, Hardie Grant Books, London.
18. M. Boelens, 1984, "Essential Oils and Aroma Chemicals from *Eucalyptus globulus*," Perfumer & Flavorist, (9): 2-14.
19. E. Vosnaki, 2020, Perfume Shrine (perfumeshrine.blogspot.com)
20. D. J. Rowe, 2005, *Chemistry and Technology of Flavors and Fragrances*, CRC Press, Boca Raton, FL.
21. G. O. Brechbill, 2009, *Perfume Bases & Fragrance Ingredients*, Fragrance Books, New Jersey (perfumerbook.com)
22. S. Arctander, 2019, *Perfume and Flavor Chemicals (Aroma Chemicals)*, Vol. I-III, Lulu Press, Morrisville, NC.
23. A. M. Rouhi, 1999, "Perfumes: A Whiff of Chemistry," C&E News, Am Chem.

Soc., Washington, D.C.
24. P. Z. Bedoukian, 1967, *Perfumery and Flavoring Synthetics*, 2nd ed., Elsevier Pub., NY.
25. W.A. Poucher et al., 1974, *Perfumes, Cosmetics and Soaps*, Wiley, NY.
26. L. Turin, 2006, *The Secret of Scent*, Harper Collins Pub., NY.
27. J. Kennedy, 2013, Table of Esters and Their Smells (jameskennedymonash.wordpress.com)
28. M. Aftel, 2001, *Essence & Alchemy: A Natural History of Perfumes*, North Point Press, NY.
29. E. T. Morris, 1984, *Fragrance: The Story of Perfume from Cleopatra to Chanel*, Scriber Pub., NY.
30. R. Genders, 1972, *A History of Scent*, Hamilton Pub., London.
31. F. Kennett, 1975, *History of Perfume*, Harrap Pub., London.
32. N. Groom, 1999, *Perfume*, Running Press, Philadelphia.
33. B. Herman, 2013, *Scent & Subversion*, Lyons Press, Guilford, CT.
34. J. Leffingwell, Leffingwell & Associates, Services and Software for the Perfume, Flavor, Food & Beverage Industries (leffingwell.com)
35. P. de Nicolai, 2008, "A Smelling Trip into the Past: The Influence of Synthetic Materials on the History of Perfumery," Chem. Biodivers. 5 (6): 1137-1146.
36. S. Delacourte, "The Composition of a Perfume" (sylvaine-delacourte.com).
37. Perfume (en.wikipedia.org).
38. J. Carles, 1961-1963, "A Method of Creation in Perfumery," Researches, 12 (11): 8-25, 1961; 12 (12): 18-29, 1962; 12 (13): 92-103, 1963.
39. M. Carles, 1974, "Education in Perfumery" in W. I. Kaufman, *Perfume*, Dutton
    Co., NY.
40. S. V. Toller & G. H. Dodd, eds., 1988, *Perfumery: The Psychology and Biology of Fragrance*, Springer-Science, Berlin.
41. L. Appell, 1982, *Cosmetics, Fragrances and Flavors*, Micelle Press, Dorset, UK.
42. L. Appell, 1997, *Formulation & Preparation of Cosmetics, Fragrances and Flavors*, Micelle Press, Dorset, UK
43. T. Curtis & D. G. Williams, 1994, *Introduction to Perfumery*, Ellis Horwood Pub., NY.
44. P. M. Müller & Lamparsky, 1994, *Perfumes: Art, Science and Technology*, Springer-Science Pub., Zurich.
45. L. Turin & T. Sanchez, 2009, *Perfumes: The A-Z Guide*, Penguin Books, London; 2011, *The Little Book of Perfumes*, Viking Press, London.
46. L. Turin & T. Sanchez, 2018, *Perfumes: The Guide*, Perfüümista ÖÜ, Tallinn, Estonia.
47. C. Burr, 2007, *The Perfect Scent*, Picador Pub., NY.
48. M. Zarzo & D. T. Stanton, 2009, "Understanding the Underlying Dimensions in Perfumers' Odor Perception Space as a Basis for Developing Meaningful Odor Maps," Attention, Perception & Psychophysics, 71 (2): 225-227.
49. J. Sparla, "Bespoke Perfumes, Perfumes that Last," (841362.wixsite.com).
50. A. Gilbert, 2008, *What the Nose Knows*, Crown Pub., NY.

51. C. Newman, 1998, "Perfume: The Art and Science of Scent," Nat. Geographic Soc., Washington, D.C.
52. P. Jellinek, 1997, *The Psychological Basis of Perfumery*, Blackie Academic & Prof., NY.
53. M. Edwards, 1992, 2008, 2011, *Fragrances of the World*, Michael Edwards & Co., Sydney, Australia.
54. L. Donna, "Fragrance Perception: Is Everything Relative," Perfumer & Flavorist, 34 (12): 26-35.
55. P. H. Pybus, 2006 "The History of Aroma Chemistry and Perfume," In: C.S. Sell, ed., *The Chemistry of Fragrances*, 2nd ed., RSC Pub., Cambridge, UK, pp.3-23.
56. C. S. Sell, 2019, *Fundamentals of Fragrance Chemistry*, Wiley-VCH, Weinheim, Germany.
57. M. Teixeira, et al., 2013, *Perfume Engineering*, Elsevier, Waltham, MA.
58. Jean-Claude Ellena, 2011, *Perfume: The Alchemy of Scent*, Arcade Pub., NY.
59. Jean-Claude Ellena, 2012, *The Diary of a Nose: A Year in the Life of a Parfumeur*, Random House, NY.
60. M. Kydd, 2007, "Exposing the Perfumer," Perfumer & Flavorist, 32: 38-44.
61. S. V. Dowthwaite, 1999, "Training the ABC's of Perfumery," Perfumer & Flavorist, May/June.
62. W. I. Kaufman, 1974, *Perfume*, E. P. Dutton & Co., NY.
63. E. Roudnitska, 1994, "The Art of Perfumery" In: P. M. Müller & Lamparsky, *Perfumes: Art, Science and Technology*, Springer-Science Pub., Zurich.
64. E. Roudnitska, 1980, "The Beginner Face to Face with the Perfumer's Palette," Perfumer & Flavorist, VIII Essential Oil Congress: 23-28.
65. E. Roudnitska, 2018, *A Life of Perfume*, Leonard Payne, England.
66. E. Roudnitska, 1974, "Creativity in Perfumery" In: W. I. Kaufman, *Perfume*, Dutton Co., NY.
67. S. B. Kaufman & C. Gregoire, 2015, *Wired to Create*, Penguin Random House, NY.
68. T. Selig, 2015, *Creativity Rules*, Harper-Collins, NY.
69. "Disclosed Chemicals Perfumes," 2010, Environmental Working Group, (ewg.org).
70. J. G. Sullivan, 2020, "Robots are Coming for Your Beauty Collection," Elle (elle.com)
71. B. Marr, 2019, Enterprise Tech, "Artificial Intelligence can now Create Perfumes" (forbes.com)
72. P. Wintermaier, 2019, Symrise and IBM: New Fragrances Created by AI" (digital.hbs.edu)
73. E.W. Kirchhoff, J. Aikens & C. Cassidy, 2000, "Formulating a Synthetic Perfume Rapidly," Chem. Innovation, 30 (11): 52-53.
74. Perfumer & Flavorist, 2018, "Firmenich Utilizes ScentMove Tech for Fragrance Creation," (perfumerflavorist.com)
75. AI: Future of Fragrance, Perfumer & Flavorist, 2/2020.
76. International Fragrance Association (ifrafragrance.org)
77. M. Bomgardner, 2019, "How Perfumers Walk the Fine Line Between Natural

and Synthetics," C&E News, 97 (16).
78. Campaign for Safe Cosmetics (safecosmetics.org)
79. Perfume Allergies, Heath & Consumers, European Commission (ec.europa.eu).
80. S. Meakins, "The Safety and Toxicology of Fragrance," In: C.S. Sell, ed., *The Chemistry of Fragrances*, 2nd ed., RSC Pub., Cambridge, UK, pp. 184-198.
81. R. Cleary, 2006, "Natural Product Analysis in the Fragrance Industry," In C. S. Sell, ed., 2006, *The Chemistry of Fragrances*, 2nd ed., RSC Pub., Cambridge, UK, pp. 214-228.
82. A. Williams, 2002, "Rose Ketones: Celebrating 30 years of Success," Perfumer & Flavorist, 27: 18-31.
83. Theugebaert, 2017, "Rose Ketones," Scent and Molecules, (scentandmolecules.wordpress.com).
84. P. K. Rout, S. N. Naik & Y. R. Rao, 2010, "Composition of Absolutes of Jasmine Sambac Flowers Fractionated with Liquid CO_2 and Methanol," J. Essent. Oil. Res., 22 (5): 398-406.
85. L.-F. Zhu, B.-Y. Lu & Y.-J. Luo, 1984, "Preliminary Study on the Chemical Constituents of *Jasminum sambac* L. Flower Fragrance," Zhiwu Xuebao, 26: 189-194.
86. X. Bu et al., 1987, "Analysis of Headspace Volatile Constituents of *Jasminum sambac* L.," Beijing Daxue Xuebao Ziran Kexueban, 6: 53-60.
87. R. Kaiser, 1988, "New Volatile Constituents of *Jasminum sambac* L.", In: *Flavors and Fragrances: A World Perspective*, B. M. Lawrence et al., eds., Elsevier Sci. Pub., Amsterdam.
88. L. Jirovetz, et al., 2007, "Chemical Composition, Olfactory Evaluation and Antimicrobial Activities of *Jasmine grandiflorum* L. Absolute from India," Nat. Prod. Communications, 2 (4): 407-412.
89. J. Coulomb, 2018, "Beyond Muget," Perfumer & Flavorist, 4/2018.
90. A. Goeke, P. Kraft & A. Alchenberger, 2018, "Discovery of Nympheal: The Definitive Muget Aldehyde," Perfumer & Flavorist, 2/2018.
91. G. Bovil, 2013, "The Future of Lavender," Perfumer & Flavorist, 38: 34-41.
92. S. Krivek, 2018, "Essential Oil Composition and Antioxidant Activities of Eight Cultivars of Lavender and Lavandin from Western Anatolia," Industrial Crops & Products, 117: 88-96.
93. P. Rakthaworn, et al., 2009, "Extraction Methods for Tuberose Oil and Their Chemical Composition," Kasetsart J. (Nat. Sci.), 43: 204-211.
94. E. M. Gaydou, R. Randriamiharisoa & J. Bianchini, 1986, "Composition of the Essential Oil of Ylang-Ylang from Madagascar," J. Agric. Food Chem., 34 (3): 481-487.
94a. M. Gautschi, M. Bajgrowicz & P. Kraft, 2001, "Fragrance Chemistry: Milestones and Perspectives," Chimia Intn'l. J. Chem, 55 (5): 379-387.
95. E.-J. Brunke, et al., 1994, "Headspace Analysis of Hyacinth Flowers," Flavour & Fragrance J., 9 (2): 59-69.
96. I. L. Bonaccorsi et al., 2011, "Composition of Egyptian Neroli Oil," Nat. Prod. Communication, 6 (7): 1009-1014.
97. Mondello et al., 1996, "Italian Citrus Petitgrain Oils. Part I. Composition of Bitter Orange Petitgrain Oil," J. Essent. Oil Res., 8: 597-609.

98. G. Dugo, 1996, "Characterization of Italian Citrus Petitgrain Oils," Perfumer & Flavorist, 21: 17-28.
99. T. A. van Beek & D. Joulain, 2017, "The Essential Oil of Patchouli, *Pogostemon cablin*: A Review," Flavour & Fragrance, 33 (3): 1-45.
100. E. Belhassen, et al., 2015, "Volatile Constituents of Vetiver: A Review," Flavor and Fragrance J., 30 (1): 26-82.
101. K. Shankaranarayana & K. Parthasarathi, 1984, "Synthetic Sandalwood Aroma Chemicals," Perfumer & Flavorist, 9: 17-25.
102. H. Hussain, et al., 2013, "Chemistry and Biology of Essential Oils of Genus *Boswellia*," Evidence-Based Complementary and Alt. Medicine, DOI: 10.1155/2014/605304.
103. L. Stern & L. Robertson, 2019, "Animal Musks: The Dark Secret of Perfume," (purrfumery.com).
104. M. Aftel, 2014, "Five Icky Animal Odors that are Prized by Perfumers," (discoverymagazine.com).
105. G. Ohloff, B. Winter & C. Fehr, 1994, "Chemical Classification and Structure Odor Relationships," In: P. M. Müller & Lamparsky, *Perfumes: Art, Science and Technology*, Springer-Science Pub., Zurich.
106. L. Turin, 2003, "Structure-Odor Relations: A Modern Perspective" in *Handbook of Olfaction and Gustation*, 2nd ed., R. L. Doty, ed., Marcel Dekker, NY.
107. L. Turin, 2002, "A Method for the Calculation of Odor Character from Molecular Structure," J. Theoretical Biology, 216: 367-385.
108. D. Anonis, 1997, "Civet and Civet Compounds," Perfumer & Flavorist, 22: 43-47.
109. A. Rinaldi, 2007, "The Scent of Life," European Molecular Biology Organ., 8 (7): 629-633.
110. P. Nef, 1998, "How We Smell: The Molecular and Cellular Basis of Olfaction," News Physiol. Sci., 13: 1-5.
111. L. Buck & R. A. Axel, 1991, "A Novel Multigene Family May Encode Odorant Receptors: A Molecular Basis for Odor Recognition," Cell, 65: 175-187.
112. L. Buck, 2005, "Unravelling the Sense of Smell (Nobel Lecture)," Angew Chem. Int. Ed., 44: 6128-6140. (nobelprize.org)
113. B. Mainic, J. Hirono, T. Sato & L. Buck, 1999, "Combinatorial Receptor Codes for Odors," Cell, 96: 713-723.
114. Z. Zhou & L. Buck, 2006, "Combinatorial Effects of Odorant Mixes on Olfactory Cortex," Science, 311: 1477-1480.
115. S. Zozulya, F. Ecgeverri & T. Nguyen, 2001, "The Human Olfactory Receptor Repertoire," Genome Biology, 2 (6): 18.1-18.12.
116. J. Amore, 1970, Molecular Basis of Odor, Charles Thomas Pub, Springfield, IL.
117. J. Amore, J. Johnston, Jr., & M. Rubin, 1964, "The Stereochemical Theory of Odor," Scientific American, 210 (2): 42-49.
118. J. Amore, 1971, *Chemical Senses*, Springer-Verlag, NY.
119. H. Boelens, 1974, "Relationships Between the Chemical Structure of

Compounds and their Olfactive Properties," Cosmetics and Perfumery, 89: 1-7.
120. K. Rossiter, 1996, "Structure-Odor Relationships," Chem. Rev., 96: 3201-3240.
121. P. Kraft, 2000, "Odds and Trends: Recent Developments in Chemistry of Odorants," Angew. Chem., Int. Ed., 39 (17): 2980-3010; 112: 3106.
122. G. Sanz et al., 2008, "Relationships Between Molecular Structure and Perceived Odor Quality of Ligands for a Human Olfactory Receptor," Chem. Senses, 33: 639-653.
123. E. R. Naipawer, et. al., 1981, "Structure-Odor Relationship for Sandalwood Aroma Chemicals," In: *Essential Oils*, E. Mookherjee & C. Mussinan, Allured Pub., Wheaton, p. 105.
124. E. R. Naipawer, 1988, "Synthetic Sandalwood Chemistry – A Decade in Review," In: *Flavors and Fragrance: A World Perspective*, B. Lawrence, et al., eds, Elsevier Sci., Amsterdam, p. 805.
125. G. Buchbauer, et al., 1994, "Conformational Parameters of the Sandalwood Odor Activity: Conformational Calculations on Sandalwood Odor," Helv. Chim. Acta, 77 (8): 2286-2296.
126. A.S. Dimoglo, et al., 1993, Dokl. Akad. Nauk SSR, Ser. Khim., 328, 570.
127. A.S. Dimoglo, et al., 1995, "Investigation of the Relationship between Sandalwood Odor and Chemical Structure: Electron Topological Approach," New. J. Chem., 19: 149.
128. A. Kovatcheva et al., 2003, "QSAR Modeling of Campholinic Derivatives with Sandalwood Odor," J. Chem. Inf. Computer Sci., 43 (1): 259-266.
129. D. Hadaruga, N. Hadaruga & S. Muresan, 2007, "Quantitative Structure-Odor Relationships of Some Nitro-Benzene Musks," Chem. Bull. Polytehnica 52 (66):1-2.
130. K. Tanaka et al., 2020, "Structure-Odor Relationships of the Main Vetiver Component of Khusimol and its Derivatives," J. Japan Assoc. Odor Envion., 51 (3): 201-204.
131. C. Sell, 2006, "On the Unpredictability of Odor," Angew. Chem. Internal. Ed., 45 (38): 6254-6261.
132. C. Sell, 2000, "Structure/Odor Correlations, the Mechanism of Olfaction and the Design of Novel Fragrance Ingredients," Perfumer & Flavorist, 25: 67 73.
133. G. Dyson, 1938, "The Scientific Basis of Odour," Chem. Ind., 57: 647-651.
134. R. H. Wright, 1977, "Odor and Molecular Vibration: Neural Coding of Olfactory Information," J. Theoretical Biology, 64 (3): 473-474.
135. C. Burr, 2004, *The Emperor of Scent*, Random House, NY.
136. S. Gane, L. Turin, et al, 2013, "Molecular Vibration-Sensing Component in Human Olfaction," Plos ONE, 8 (1): e55870.doi:10.1371/journal.pone.0055780.
137. R. C. Jaklevic & J. Lambe, 1966, Molecular Vibration Spectra by Electron Tunneling," Phys. Rev. Letters, 17 (2): 1139.
138. J. C. Brookes, et al., 2007, "Could Humans Recognize Odor by Phonon Assisted Tunneling?" Physical Review Letters, 98.

139. M. Franco, et al., 2011, "Molecular Vibration-Sensing Component in *Drosophila melanogaster* Olfaction," Proc. Nat. Academy of Science, 108 (9): 3797-3802.
140. J. C. Brookes, A. Horsfield & A. M. Stoneham, 2008, "Odor Characteristics for Enantiomers Correlate with Molecular Flexibility," J. Royal Soc. (doi.org/10.1098/rsif2008.0165)
141. R. Bentley, 2006, "The Nose as a Stereochemist: Enantiomers and Odor," Chem. Rev., 106: 4099-4112.
142. E. Block, et al., 2015, "Implausibility of the Vibrational Theory of Olfaction," PNAS, pnas.org/cgi/doi/10.1073/pnas.150305112.
143. A. Keller & L. B. Vosshall, 2004, "A Psychophysical test of the Vibration Theory of Olfaction," Nature Neuroscience 7 (4): 337-338.
144. A. Keller & L. B. Vosshall, 2016, "Olfactory Perception of Chemically Diverse Molecules," BMC Neuroscience 17 (1): 55.
145. L. B. Vosshall, 2015, "Laying a Controversial Smell Theory to Rest," Proc. Nat. Acad. Sci., 112 (21): 6525-6526.
146. I. A. Solov'you, P.-Y. Chang & K. Schulten, 2012, "Vibrationally Assisted Electron Transfer Mechanism of Olfaction: Myth or Reality," Physical Chem.-Chem. Physics, DOI:10: 1039/C2CP41436H.
147. F. Kermen, et al., 2011, "Molecular Complexity Determines the Number of Olfactory Notes and the Pleasantness of Smells," Scientific Reports, 1: 206.
148. J. C. Brookes, 2010, "Science is Perception," Phil. Trans. R. Soc., 368: 3491 3502.
149. J. C. Brookes, A. Horsfield & A. M. Stoneham, 2012, "The Swipe Card Model of Odorant Recognition Sensors," 12: 15709-1574.

Appendix I. Odors of Some Perfumery Compounds

Aldehydes

Acetaldehyde	Ethereal
Acetophenone	Pungent, orange
Aldehyde C6	Green, Grassy (Hexanal)
Aldehyde C7	Fatty, Green, Herbaceous (Heptanal)
Aldehyde C8	Orange (Octanal)
Aldehyde C9	Rose (Nonal)
Aldehyde C10	Orange rind (Decanal)
Aldehyde C11	Clean, Citrus-Floral (Undecanal)
Aldehyde C12	Soapy, Lilacs/Violet (Lauryl aldehyde)
Aldehyde C13	Waxy Grapefruit (Tridecanal)
Aldehyde C14 & C18 are lactones	
Aldehyde C16 is an ester	
Amyl-cinnamic aldehyde	Jasmine
Anisic aldehyde	Floral, Sweet, Hawthorne
Benzaldehyde	Almond-like
Bourgeonal	Muget, Floral, Watery Green
Cortexal	Leafy-Green (Homo-cuminic aldehyde)
Cyclamen aldehyde	Floral green, watermelon, fresh rhubarb
Floralozone	Clean, Green, Fresh Air-Ocean Breeze
Furfural	Burnt oats
Helional	Fresh, Watery green, Metal-cold air, Hay
Heliotropin	Floral, Mimosa, Vanilla-cherry (Piperonal)
cis-3-Hexenal	Grassy, Green tomatoes
Hexyl-cinnamic aldehyde	Jasmine
Isovaleraldehyde	Nutty, Fruity, Cocoa
Lilial	Muget
Methyl-nonyl acetaldehyde	Herbal-pine (MNA)
Methyl-octylacetaldehyde	Orange (MOA)
Muget aldehyde	Muget, Rose (citronellyl oxyacetaldehyde)
2,6-Nonadienal	Violet-green
Nympheal	Muget
Oncidal	Green
Phenylacetaldehyde	Green
α-Sinensal	Citrus, Orange, Mandarin
Syringa aldehyde	Green
Triplal	Green Leafy, Tomato

Ketones

Ambretone	Musk, Sweet, Floral
Calone	Watery, Fresh, Sea Breeze (Watermelon ketone)
D-Carvone	Spicy, caraway
R-Carvone	Sweet, spearmint
Cashmeran	Woody, Musk, Amber, Leather
Civetone	Strong musk
Cosmone	Musk, Powdery, Ambergris-like
Damascenone	Sweet, Fruity, Rose-plum
α-Damascone	Rose, Apple
β-Damascone	Rose, Black currant, Plum
Dihydrojasmone	Fruity, Woody, Floral
Exaltone	Musk
Fructone (a Ketal)	Apple with Strawberry & Woody aspects (a Ketal)
Hedione	Fresh, Floral, Jasmine
α-Ionone	Floral, Violet
β-Ionone	Floral, Woody, Sweet, Berry
Irone	Floral, Iris, Woody
Iso E Super	Smooth, Woody, Amber
Methyl amyl ketone	Spicy
Methyl-dihydrojasmonate	Fresh, Floral, Jasmine (Hedione)
Muscone	Musk
Nootkatone	Grapefruit
Phantolide	Musk
Tagatone	Herbaceous
Tonalid	Musk Ketone
Veramoss	Mossy, Oakmoss, Woody (Evernyl)
Versalide	Musk
Vertofix	Woody, Vetiver, Leather (methyl cedryl ketone)
α-Vetivone	Woody, Balsamic, Floral

Ether

Ambrox	Sweet, Musky, Ambergris-like
Benzyl methylether	Fruity, Ethereal
Galoxolide	Clean Musk, Woody, Floral
Methyl cedryl ether	Ambergris-like, Woody

Lactones

Amberketal	Ambery, Woody, Dry
Ambrettolide	Musky, Floral, Sweet
Coumarin	New mown grass, Hay w/vanilla scent
γ-Decalactone	Creamy, Peach
Exaltolide	Sweet, Musky-animal
Habanolide	Musk, Powdery, metallic-hot iron

Jasmine lactone	Fatty, Fruity, Peach/Apricot
Massoia lactone	Creamy coconut
γ-Nonalactone	Coconut
δ-Octalactone	Creamy
Paradisone	Intense Jasmine
Prunolide	Creamy, Coconut (Aldehyde C18)
Spirambrene	Woody, Amber, Spicy

Alcohols

Ambermax	Amber, Woody
Benzyl alcohol	Bland, sweet floral, damp
Benzyl isoeugenol	Balsamic
Cedrol	Woody
Cinnamic alcohol	Balsamic
Citronellol	Citrus, Clean, Rosy
Dihydromyrcenol	Fresh, Clean, Lime, Lavender
Ethyl Maltol	Caramelized sugar, cotton candy (Veltol)
Farnesol	Fresh Floral
Furaneol	Strawberry
Geraniol	Rose
1-Hexanol	Herbaceous, Woody
cis-3-Hexen-1-ol	Fresh cut grass
Hydroxycitronellol	Floral, Muget, Rose
Khusimol	Woody, Earthy
Javanol	Woody, Milky, Sandalwood
Linalool	Floral, Fruity
Methyl diantilis	Spicy, Powdery, Smoky
Nerolidol	Woody, Fresh Bar
Orcinyl-3	Oakmoss, Fruity, Marine
Osyrol	Woody, Milky, Sandalwood
Patchouli alcohol	Woody. Musky
Phenylethyl alcohol	Floral, Sweet Rose
Rhodinol	Fresh Rose
Santaliff	Woody, Milky, Sandalwood
Santalol	Woody, Milky, Sandalwood
α-Terpineol	Lilac, Citrus, Pine
γ-Terpineol	Herbaceous, Terpy, Citrus
Undecylenic alcohol	Soapy, waxy, floral, rosy pleasant
Vetiverol	Woody, earthy

Esters

Ally amyl glycolate	Fruity, Green, Galbanum
Amyl acetate	Fruity, Banana
Amyl salicylate	Herbaceous, Floral, Azalea
Benzyl acetate	Floral, Jasmine, Fruity
Benzyl benzoate	Floral, Spicy, Aromatic
Benzyl salicylate	Fresh Azalea, Daffodil
Bornyl acetate	Pine
Cedryl acetate	Woody-Cedar, Leather, Balsamic
Cyclogalbanate	Herbal-Green, Fruity, Galbanum
Cyclomusk	Musk
Ethyl acetate	Sweet, Fruity
Ethyl butyrate	Fruity, Orange, Pineapple
Ethyl cinnamate	Sweet, Oriental
Ethyl-2-methoxybenzoate	Sweet, Floral, Fruity
Ethyl-methylphenylglycidate	Strawberry
Ethylene brassylate	Musk, Powdery, Ambrette-like
Evernyl	Musky, Oakmoss (Veramoss)
Fructone	Fruity, Apple
Geranyl acetate	Fruity, Rose
Helvetolide	Musk, Floral, Fruity
3-Hexenyl tiglate	Green, Earthy, Floral
Hexyl acetate	Apple, Floral, Fruity
Isoamyl acetate	Fruity, Banana, Pear
Linalyl acetate	Lavender-Bergamot
Methyl benzoate	Strong Floral
Linalyl cinnamate	Woody, Floral, Green
Methyl acetate	Sweet, Nail Polish (acetone)
Methyl anthranilate	Fruity, Grape
Methyl butyrate	Fruity, Apple, Pineapple
Methyl-dihydrojasmonate	Fresh, Floral, Jasmine (Hedione)
Methyl formate	Ethereal
Methyl-heptine carbonate	Fresh Green Vegetable, Violet (MHC)
Methyl-octane carbonate	Violet leaf, Green (methyl octyonate)
Methyl salicylate	Minty, Spicy wood
Methyl propionate	Sweet, Fruit, Rum
Octyl acetate	Fruity, Orange
Paracresyl acetate	Floral, Narcissus, Animalic
Pentyl pentanoate	Fruity, Apple
Phenylethyl acetate	Rose, Honey

Phenylethyl-phenyl acetate	Heavy sweet Floral, Balsamic
n-Propyl acetate	Pears
PTBCHA	Woody, Orris, Creamy (polytertbutyl cyclohexyl acetate)
Romandolide	Musk, Ambrette-like
Styrallyl acetate	Green floral, Gardenia
Vetiveryl acetate	Fresh Woody, Sweet, Amber

Linear & Branched Terpenes

Citral, Geranial	Lemon
Citronellal	Citrus, Lemon, Floral
Citronellol	Lemon
Galbanolene	Greenish Metallic, Galbanum
Geraniol	Rose, Flowery
Hydroxycitronellal	Floral, Muget
Linalool	Floral, Sweet, Woody-Lavender
Myrcene	Woody
Nerol	Sweet, Rose, Flowery
Nerolidol	Wood
Ocimene	Tropical green, Woody, Floral
β-Pinene	Woody, Green, Pine
α-Thujene	Woody, Green, Herbal
Valencene	Citrus

Cyclic Terpenes

α-Bulnescene	Woody
Camphor	Camphoraceous
Carvone	Caraway or Mint, dependent on stereoisomer
Caryophyllene	Soft Spicy, Woody
Eucalyptol	Camphoraceous, Minty, Cooling
α-Guaiene	Sweet, Woody, Peppery, Balsamic
α-Ionone	Violet, Woody
Limonene	Orange
Terpinene	Woody, Terpy, Citrus
Terpineol	Lilac, Citrus
Thujone	Minty

Aromatic Terpenes

Anethole	Anise
Benzaldehyde	Almond
Cinnamaldehyde	Cinnamon
Ethyl maltol	Caramelized sugar, cotton candy
Eugenol	Clove
Vanillin	Sweet, vanilla, chocolate, balsamic

Amines, Nitrogen Containing

Cadaverine	Rotting Flesh
Fleuranil	Fresh, Floral, Ozone, Salted Odor
Galbanum pyrazine	Bell Pepper, Peas, Galbanum
Galbazine	Green Pepper, Galbanum
Indole	Fecal, Flowery
Isobutylquinoline	Woody, Leather, Vetiver-like
Neocaspirene	Floral, Black Currant Bud
Pyridine	Fishy
Skatole	Fecal

Appendix II. YouTube Lectures (URLs) On Perfumes and Theories of Smell Raymond Young "Cultural Education" Channel

Essential Oils & Perfumes
Part I. Sources and Collection of Essential Oils
  (https://youtu.be/d-QXnXOg1dU)
Part II. Characteristics of Essential Oils (Blossoms)
  (https://youtu.be/wqZdYMuDBMc)
Part III. Characteristics of Essential Oils (Other Plant Sources)
  (https://youtu.be/g6_k3Wo4v1A)
Part IV. Characteristics of Odorants from Animal Sources
  (https://youtu.be/DdW0zbpqTHg)
Part V. Perfumes: Structure & Composition
  (https://youtu.be/tcj0QNsVGyo)
Part VI. Perfume Accords (Families)
  (https://youtu.be/MPdcVQbL_oY)
Part VII. Perfume Accords (Families) Modern & Summary
  (https://youtu.be/N6PvWekt124)

Theories of Smell
Part I – Introduction to Chemistry of Odorants
  (https://youtu.be/uzJaRAsey-8)
Part II – Chemistry of Natural Odorants
  (https://youtu.be/4AlS-urHud0)
Part III –Introduction to Perfumes & Physiology of Smell
  (https://youtu.be/9X6wmKLalnM)
Part IV –Molecular Shape Theories of Smell
  (https://youtu.be/83eXb6Xdyfw)
Part V –Vibrational Theories of Smell
  (https://youtu.be/_1xIqu-ltd0)
Part VI –Combinatorial Odotope Theory of Smell & Summary
  (https://youtu.be/V4uzyZ_vFJo)

About the Author

I am an Emeritus Professor from the University of Wisconsin-Madison where I specialized in Natural Polymer Chemistry & Ethnobotany. My B.S. & M.S. degrees are from Syracuse University and my Ph.D. from the University of Washington-Seattle. I was a Fulbright Scholar/American Scandinavian Fellow as a Graduate Student at the Royal Institute of Technology in Stockholm, Sweden and a Senior Fulbright Scholar as a Professor at the Aristotle University of Thessaloniki, Greece. I have published many original scientific papers, book chapters & patents and have edited or authored 8 books, further information is at:
https://www.researchgate.net/profile/RA_Young

Subject Index

A
Acetone, 92, 93
Alcohols, 37, 109, 181
Aldehydes, 36, 44, 58, 61, 67, 83, 110, 118, 135, 179
Aliphatic hydrocarbons, 7, 106
Amber note, 20, 30, 43, 47, 67, 70, 73
Ambergris, 30, 63, 72, 125, 152
Ambrette, 21
Ambrox, 145, 152
Amyl salicylate, 62, 72
Animal odorants, 7, 15, 24, 30, 47, 89 125, 148
Anise, 11, 21
Anosmia, 90
Apple, 23, 45, 52, 67, 140
Aromatherapy, 7, 16, 18-20, 22-26
Aromatic hydrocarbons, 108
Aromatic odor, 51, 54, 64, 65

B
Balsamic note, 9, 22, 39, 55
Bases, 44, 84-86
Base Note, 42, 46-53, 82-84, 93
Beaux, Ernest, 59
Benzene, 108
Benzoic acid, 28
Benzoin, 27, 29, 46,7 0
Benzyl acetate, 16, 37, 115, 128
Benzyl benzoate, 17, 130, 135, 139
Bergamot, 12, 15, 23, 60, 69
Birch, 90
Bitter Orange, 12,1 8, 93,1 37
Black Currant, 23, 52, 70
Buchu,18, 24
Buck, Linda, 167

C
Calone, 37, 51, 66-70, 90, 113
Camphor, 11, 19, 27, 49, 130, 158
Caramel, 20, 22, 67, 69, 72, 141
Caraway, 116
Cardamon, 21, 22

Carles, Jean, 42, 77-79, 81-85, 90, 92
Carnation, 61, 74, 84, 135
Carvones, 35, 117
Cashmeran, 151
Castoreum, 32, 61
Cedarwood, 11, 25, 40, 43
Cedrol, 27, 109, 117
Cedryl acetate, 62
Chanel, Coco, 48
Chemical classification, 117, 179, 34, 106
CO_2 extraction, 11, 14, 22, 24, 127
Cinnamaldehyde, 35
Cinnamic acid, 28
Cinnamic alcohol, 127, 136
Cinnamon, 18, 25, 40, 43, 61
Citral, 18, 116, 117
Citronellal, 19, 35, 37, 116
Citronellol, 15, 19, 35, 90, 116, 121, 128, 134
Citronellyl acetate, 121
Citrus, 15, 23, 35, 42, 50, 66, 68, 139
Civet, 32, 112, 148
Covetone, 112, 149
Computer technology, 98
Concretes, 11, 127
Costus, 20, 21
Coumarin, 21, 46, 58, 90, 112, 141, 170
Cyclomusk, 152

D
Damascenone, 65,1 23
Damascones, 37, 64, 96, 112, 123
Damask rose, 14, 122, 123
Dihydrojasmonate, 34, 63, 67
Dihydromyrcenol, 66, 109

E
Earthy note, 20, 29, 33,5 5, 62, 142
Electron tunneling, 164
Ellena, Jean-Claude, 77, 87, 91-94, 135

Enantiomers, 116, 166
Enfleurage, 8, 125, 131
Esters, 37, 38, 113, 116, 181
Ethers, 180
Ethyl maltol, 37, 67, 109, 141
Ethylene brassylate, 158
Eucalyptus, 8, 11, 19
Eugenol, 109, 117, 135, 170
Evernyl, 46, 63, 145, 147
Exaltolide, 150
Exaltone, 149
Expression, 8, 12, 139

F
Facet, 44
Farnesenes, 117
Farnesol, 109, 117, 127
Fig Leaf, 92
Floral, 20, 26, 29, 40, 47-54, 60, 71, 89, 122-138
Frankincense, 11, 27, 40, 144
Fresh Green Note, 47, 50, 54, 145
Fruity, 23, 26, 29, 43, 52, 54, 69, 71, 89, 131,140

G
G-Protein, 7-fold helix, 155
Galaxolide, 67, 97, 151
Galbanum, 30, 63, 71, 91, 146
Gardenia, 43, 71, 135
Gas chromatography, 34, 119
Geraniol, 15, 35, 37, 109, 114, 116, 127
Geranium, 11, 15, 18, 35, 64, 134
Geranyl acetate, 114, 132
Grapefruit, 23, 35, 92, 111
Grasse, 8, 77
Green odor, 47, 51, 54, 63, 111
Grojsman, Sophia, 65, 66, 96, 97

H
Hawthorne, 68
Headspace, 120, 126, 135-137
Health aspects, 102, 156
Heart note, 42, 46-53, 73, 82-84, 93
Hedione, 63, 68, 97, 112, 117,125, 151
Helional, 73, 118
Heliotrope, 14, 41, 71
Heliotropin, 59, 68, 90
Helvetolide, 152
Herbaceous note, 55, 64, 135
Honey, 26, 33, 70
Hyacinth, 14, 41, 136
Hydroxycitronellal, 128

I
Incense, 40
Indole, 16, 92, 117
Ionones, 37, 46, 58, 96, 112, 116, 123, 127, 133, 151
Iralia base, 133
Iris, 14, 20, 26, 74, 138
Irones, 35, 113, 138
Iso E Super, 67, 90, 97, 151
Isomers, 116, 166
Isoraldeine, 133

J
Jasmine, 15, 35, 43, 73, 96, 125
Jasmone, 16, 125, 130
Javanol, 144, 159

K
Ketones, 36, 117,1 80
Khusimol, 21, 142

L
Labdanum, 27, 29, 44, 70, 145
Lavandin, 16, 52, 129
Lavender, 8, 11, 16, 19, 42, 129
Leather, 30, 50, 70
Lemon, 12, 18, 23, 35, 139
Lemon grass, 18, 35
Lilac, 14, 97, 134
Lilial, 118, 127, 163
Lily, 14
Lily-of-the-Valley, 14, 26, 41, 62, 65, 73, 127, 134
Lime, 12, 23, 35, 51, 140
Limonene, 28, 115, 117, 144,
Linalool, 16, 26, 35, 37, 90,1 09, 114,

116, 131, 135, 139, 144
Linalyl acetate, 17, 23, 35, 52, 83, 114, 117, 130, 137, 139
Lyral, 111, 118, 127

M
Methyl anthranilate, 17, 93, 118, 125, 130, 135
Mimosa, 41
Molecular modeling, 160
Molecular shape, 157-164
Molecular weight, 154
Monoterpenes, 115
Muget, 14, 43, 59, 68, 84, 97, 127, 163
Muscone, 33, 113, 149
Musk
  ambrette, 148
  Bauer, 148
  deer, 32
  ketone, 61, 82, 148
  xylene, 148
Musks, 20, 31-33, 43, 46, 66, 68, 82, 92, 102, 118, 148-152, 154
  dione, 152
  linear, 151
  macrocyclic, 68, 149, 162
  polycyclic, 150
Myrcene, 144
Myrrh, 27, 40, 64

N
Narcissus, 41
Nerol, 15, 70
Neroli oil, 11, 13, 137
Nitro musks, 148, 162

O
Oakmoss, 30, 43, 71, 82, 96, 102, 147
Ocimene, 130, 134, 144
Odor
  description, 53, 55, 124
  measurement, 124
  threshold, 124
Odorant receptor, 154, 167
Olfactory region, 154

Olfactory training, 77-88
Olibanum, 11, 27, 44, 63
Opopanax, 27, 65, 93
Orange blossom, 8, 13, 91
Orange oil, 12, 23, 35, 111, 119, 139
Orcinyl, 46
Orris, 14, 20, 35, 63-65, 138
Osyrol, 144, 159

P
Paradisone, 91
Passionfruit, 69
Patchouli, 8, 18, 43, 46, 62, 69, 142
Patchoulol, 20, 142
Peach odor, 23, 43, 46, 60, 69, 71, 97, 111
Perfume
  composing, 88
  blending, 89-95
  horizontal, 96
  overdosing, 96
  reactions, 118
  stability, 118
  structure, 42
Peru Balsam, 28
Petitgrain, 13, 137
Phenol, 30
Pine, 11, 18
Pineapple, 23, 52, 140
Pinenes, 19, 24, 28, 139, 144,
PTBCHA, 63

Q
Quinolines, 50, 61, 90

R
Raspberry, 69, 74, 140
Resins, 12, 27
Resinoids, 8, 12, 46, 55
Rhodinol, 91
Rose, 11, 14, 40, 43, 66, 73, 80-82, 111, 122
Rose ketones, 123
Rose oxides, 123
Rosewood, 25, 26, 64
Roudnitska, Edmund, 34, 52, 62, 77,

79-81, 90

S
Safety aspects, 102, 156
Sandalwood, 11, 25, 43, 60, 65, 72, 92, 143, 158
Santaliff, 144, 159
Santalols, 26, 143, 159
Schiff base, 118
Smell theories
    combinatorial, 167, 185
    molecular shape, 157, 165, 169, 185
    swipe-card, 170
    vibrational, 165, 185
Spicy note, 26, 30, 43, 47, 89
Steam distillation, 8, 17-19, 23, 28, 40
Storax, 29
Strawberry, 67, 140
Styrax, 29, 35, 44, 67, 119
Synthetics, 7, 34-38, 41, 53, 57, 94, 129

T
Tagatone, 145
Terpenes, 37, 183
Terpineols, 21, 24, 26, 35, 115, 128
Thiols, 24, 163, 165
Thujene, 28, 144

Tobacco, 50
Tolu balsam, 28
Tonka bean, 21, 70, 72, 112
Top note, 42, 46-53, 82-84, 93
Tuberose, 17, 65, 130

U
Un Jardin Sur Le Nile, 92
Undecalactone, 46, 58, 111, 117

V
Vanilla, 7, 21, 35, 43, 70, 72
Vanillin, 22, 29, 35, 58
Vapor pressure, 44, 81, 154
Veramoss, 63, 147
Versalide, 151
Vetiver, 20, 35, 46, 62, 83, 90-142
Vetiveryl acetate, 44, 46
Vetivone, 21, 142
Violet, 14, 18, 41, 60, 90, 111, 137
Volatility, 44, 81, 154

W
Woody note, 20, 26, 29, 47, 49, 54, 67, 72, 89, 92, 142

Y
Ylang-Ylang, 8, 17, 43, 46, 62, 131

www.ingramcontent.com/pod-product-compliance
Lightning Source LLC
Chambersburg PA
CBHW060413220526
45465CB00008B/2869